HOW TO TALK TO THE BIRDS AND THE BEASTS

HOW TO TALK TO THE BIRDS AND THE BEASTS

By Their Body Language & Their Sounds You Shall Know Them

by Jacques LeComte and
Dorothée Koechlin-Schwartz
in Collaboration with Hugues de Bonardi
Translated by Eileen Finletter
Illustrations by Harriet Springer

ARBOR HOUSE New York

Published by arrangement with Ram Publishing Company, Ltd.

Published in the United States of America by
Arbor House Publishing Company
and in Canada by Fitzhenry & Whiteside, Ltd.
Library of Congress Catalog Card Number: 79-54013
ISBN: 0-87795- 252-3
Manufactured in the United States of America
10 9 8 7 6 5 4 3 2 1

To Jean-Claude Ibert

Contents

"You have received Nature as a gift,
Know how to respect it
And transmit it to your children.
In this way from dark you will pass into light
You will walk in the brightness of day
Rivers of water will gush out
The mountains will cry out with joy
The peace of God will return to your heart
You will be worthy of doves
And the doves will envy you!"

EXTRACTS FROM THE BIBLE
BY GEORGES NEVLIP

Introduction

A wolf howls in the Alaskan night, a bee traces circles at the entrance to its hive, a male pigeon puffs out its breast, strutting and dancing for its mate. Communication among animals has been found to be incredibly rich and varied. All the ordinary senses are used—sight, hearing, smell, taste, touch—as well as some extraordinary senses we are only just beginning to understand, such as the ability of sharks and other fish to sense electrical fields. But one thing is no longer in any doubt: All animals—even the fiercest and most solitary—do communicate with their parents, mates, offspring and occasionally with outsiders, whether friend or foe. Communication is essential to social contact and often to survival.

For many years, researchers have been accumulating data on animal communication and publishing their findings in scientific journals and monographs. Unfortunately, for most of us this material is either inaccessible or completely impenetrable. Amateur naturalists and others have had to depend almost completely on their own powers of observation in learning about animal communication. It is our hope that this book will fill the gap between what is known to the few and what is understood by all. Fuller knowledge of animals and their languages can be immensely satisfying. We may not yet be able to "talk" directly with animals, but we can at least interpret their signals and codes and understand what they wish to convey.

Throughout the book, we have tried to avoid scientific jargon while still offering the reader precise and accurate information. We have translated into human language what animals are saying about their motivations and psychological states. This does not mean that the book can be used as a dictionary of animal languages. It must be remembered that an animal's signals can be meaningless when taken out of context. For example, a bird's song is part of an ecological whole and can be interpreted as a love song only when we also know something about its physiological state at the time. Some birds also change color or appearance during mating, or occupy a specific territory, and there can be any number of other variables which will change the meaning of the song. In some cases the same song or display can serve either as a sign of aggression or a call to mate. Correct interpretation will depend on careful observation of circumstances and the environment; once we isolate the song or other signal, we are liable to misinterpret it. Only when we also understand something of the animal's range of signals and habits can we interpret its language with any accuracy.

Behaviorists and other scientists are not the only ones who have come to recognize the importance of understanding the context of animal communication. Hunters and breeders who have worked with animals for many years, and amateur naturalists who have developed their observational skills, have also accumulated a wealth of knowledge. Some have written books and others are willing to talk to anyone who is interested. They should be listened to attentively.

As most people familiar with animals must know, this book is perforce anything but complete. It is merely a guide —an introduction—offering information to arouse the interest and, we hope, excitement of its readers. If it inspires even one of them to explore further and to try to answer some of the riddles of communication which are still puzzling scientists, we will have succeeded beyond our expectations.

Some Basic Ideas of Ethology

Ethology is a comparatively new science. It is the study of animal behavior with special emphasis on evolutionary adaptations to the natural world. We cannot discuss animal communication without also discussing animal behavior. Briefly, we will review some of the general ideas of ethology by answering a few of the most frequently asked questions about it.

ARE AN ANIMAL'S MEANS OF COMMUNICATION ACQUIRED BY EXPERIENCE OR ARE THEY INNATE? There is no simple answer to this question. Consider the case of the cuckoo. Its mother lays her eggs secretly in the nests of other birds. The young cuckoo is then raised by another bird species whose song differs totally from its own. Yet when the cuckoo matures, it sings a proper cuckoo song, even though it has never heard one from its foster parents. In another example, a pigeon raised without its parents coos normally once it becomes an adult. In both cases the character of the songs appears to be inherited rather than learned.

On the other hand, a bullfinch can only learn to sing by listening—not merely to another bullfinch but to its parents. A bullfinch raised by canaries will eventually sing the canary's song even if there is another bullfinch singing in a neighboring cage. It must learn its song from the one who provides its food, in this case its foster parents. The bullfinch's song owes little to heredity.

Between the two extremes, we find many species whose songs consist of some elements which are instinctive and others which must be learned. In fact there are actually bird dialects peculiar to a specific region, even to a single valley. These dialects are passed from generation to generation. The young imitate the songs of the adults, and because they usually return to their birthplace after migration the survival of the dialect is assured.

Occasionally birds of the same species but with different

dialects will not be able to understand each other, just as Yorkshire English may be incomprehensible to someone from California. Some animals can adapt to the different dialects they encounter during the course of their wanderings. This is true of the elephant seal whose grunting pitch varies from island to island but who quickly adopts the local accent when one is encountered.

So there is ample evidence that virtually no action is completely innate. In some way all behavior is modified by experience. Yet experience alone is not enough for the acquisition of a kind of behavior either. In most cases, behavior is the result of the innate *and* the learned. Both play a part in varying degree from species to species, and from case to case.

WHAT IS MEANT BY AN ANIMAL'S TERRITORY? Many animals roam about considerably. During the course of a year, these trips can cover anywhere from a few meters to several thousand miles. Some sea birds migrate to and from the North and South poles *every year.* The widest area over which an animal ranges during its life cycle is its natural habitat. Within this area there exists a smaller, "privileged" zone claimed by one animal (or group) for its own. This is the animal's territory, which it will defend against intrusion by others of the same species, except for a current or prospective mate. Defense does not necessarily mean combat in every instance. It can mean simply warning others away through use of a signal understood by the members of its own species.

Territories can be the property of an individual animal, family or group. They can be clearly delineated areas, as in the case of the stickleback (a small pugnacious freshwater fish), or they can be simply lines of communication—"boulevards"—as in the case of the large cats. Some territories are extremely small—a few square inches for the cricket—and others can be vast—an entire mountain valley reserved for an eagle and its mate. Boundaries can be identified by the behavior of the occupant or by static "frontier posts." The bear claws tree trunks; the European bison tears tree bark, coating the fresh scars with a mixture of mud and urine; the antelope deposits a secretion from its tear gland; the cat or dog leaves drops of urine at given points. Scent-marking and such signals will deter most intruders with a net saving in energy—a major priority of all biological systems.

Territories, however, are not established solely to satisfy the animal's instinct for property. Their functions are varied. Some territories serve only long enough to shelter the occupants during mating, while others are used for rearing the young. But there is more to it than function alone. Territorial rights appear to be fundamental to an animal's nature. Animals which are prevented from acquiring a territory of adequate size are often unable to behave normally. For example, when there is severe competition for space, males without territory of their own are denied access to females and unable to reproduce. This natural population control is found among animals as diverse as fish, birds, mammals and amphibians such as frogs.

The territorial instinct increases the chances of survival of a species by making the best distribution of individuals throughout the available space, thus avoiding excessive competition or the depletion of natural resources. Territory also means added protection against predators, since familiarity with the terrain makes escape easier. The animal is more alert to intruders and more aggressive in its own defense, on its home ground.

WHAT IS SOCIAL HIERARCHY, AND WHY DO SOME ANIMALS DOMINATE WHILE OTHERS ALLOW THEMSELVES TO BE DOMINATED? The phenomenon of dominance was first recorded and analyzed by researchers a little over fifty years ago. It had been observed long before that, however, by shepherds, cowherds, dog-sled guides and others who had close contact with animals in groups. They noticed that some animals laid down the law while others obeyed. The first scientific study was conducted within a henhouse. There, researchers found a rigid hierarchy—the pecking order. A dominant hen could peck a hen lower in the order without being pecked in return. Of course, among other animal societies, dominance is expressed in many other ways —right of unhindered passage, exclusive possession of females, first access to food, and so on.

In observing certain groups, it is important not to confuse truly dominant members with those who perform special tasks, such as "soldiers" or "guide" animals. For example, in a herd of cows in transit, the guides are only middle-ranked cows, but they precede the others. The dominant animals follow in the center of the group, with the lower ranked animals bringing up the rear. Similar observations have been made about other animals, such as monkeys. Nor should dominance be confused with aggressiveness. In many groups a lower ranked animal will exhibit the greatest aggressiveness. The hierarchies among groups of mammals —herds of cows, troops of monkeys and so on—are usually more complex than the linear orders most often noted among flocks of birds. Among mammals, the hierarchies are often triangular. One animal dominates another who dominates a third. Whereupon, in a sort of Möbius strip of animal hierarchy, the third animal completes the triangle by dominating the first animal. Confusing, perhaps, but true. And if this triangle is composed of strong individuals, it may dominate another such triangle composed of weaker individuals.

Under normal conditions in the habitat, hierarchies are maintained with a minimum of clashes. In the normal course of hierarchical evolution, the young progressively replace their elders. Occasionally, puberty is accompanied by a disregard of the hierarchy, an increase in aggressiveness and open combat, but such occasions are rare. In most cases, the hierarchy establishes and maintains itself with the aid of signals which can sometimes be very subtle.

Like territory, dominance and hierarchical structures are energy-conserving systems. They prevent internal struggles and assure the cohesion and greater efficiency of the group.

Most animals seem to fall into one of two categories: those who maintain strict observance of territorial rights and those with a very rigid hierarchical system. With some exceptions, hierarchic systems predominate in harsh environments where survival is easier in a group—safety in numbers. On the other hand, more satisfactory environments seem to encourage systems founded on individual territorial rights, although as we will see one must not be too quick to generalize.

WHAT DO ANIMALS COMMUNICATE TO EACH OTHER? The answer centers on the most important aspect of animal life because, as far as we can tell, most animal communication is related to survival—whether of the individual or the species.

The rituals of courtship and mating are fundamental to all animal life and the signals relating to them are extremely varied. For some species, the language is simple, such as the odor of the night moth; for others it is complex, as in the nuptial dances of certain birds. In many cases *love signals* and displays resemble those used to disarm aggression—*signals of appeasement.* This is most often true among animals who normally avoid one another, coming together only to mate and assure the survival of their species.

The signals of appeasement are especially important for animals which live in groups—wolf packs, monkey troops, deer herds and so on—and in relations between the young and their elders. These signals are an integral part of any hierarchic system and are quite varied. The less dominant animal often adopts a totally submissive attitude, placing itself at the mercy of the dominant animal—the wolf offers its throat; the stork turns its head to the side, rendering its bill harmless.

In some cases, the weaker animal offers nourishment, real or symbolic, to the dominant one. Often submission takes the form of body contact. Grooming is probably the best known form of this. Monkeys, chimpanzees, gorillas—in fact most primates—groom each other for fleas, ticks or other parasites. This is a symbolic behavior, for in many cases the dominant animal being groomed is actually free of parasites.

Signals of aggression are made before combat. In some cases, the threat display itself will be enough to drive off an adversary. The ultrasonic cries of rodents, the raised ears of squirrels, the chest pounding of male gorillas and so on are all signals of aggression.

The *signals of danger* and distress, such as the whistling of marmots or the cries of numerous birds, are most often vocal—or perhaps one should say "audible," so as not to exclude the alarm of the cricket's legs or the slap of the beaver's tail. Ornithologists have found that there are universal cries of danger understood by birds of several different species which will band together in a "coalition," temporarily setting aside interspecies rivalry to ward off a common enemy such as a cat, fox or other predator.

Some animals warn others of danger by excreting a chemical substance which provokes flight among members of the same species. Mollusks, fish, tadpoles and others use this signal of danger.

Among certain animals there are also *working signals* which are still more complex than those already mentioned. They alert the group to the need for a cooperative group effort. For example, when an individual bee has found a sizable food supply, it returns to its hive and performs a dance which communicates the exact location of the supply to the other members of the hive. For the same reason, ants excrete substances to mark their trails, and kitty hawks use special cries.

Behavioral studies are only beginning to unravel some of the mysteries of animal communication. The animal world still conceals much from its human visitors. Animals are able to sense things which we cannot perceive or understand. For example, many animals have been observed to react to an impending earthquake hours before man's most sensitive instruments begin to register the danger. What signals or vibrations can they receive from the earth? Many less obvious signals no doubt escape us as well and remain shrouded in mystery, waiting to be discovered—perhaps only with the help of animals themselves.

Bird Songs and Animal Cries

No present method of transcribing animal and bird sounds is completely satisfactory. If the signals are short they can simply be spelled phonetically—"tic," "woof," "krak," etc.—but complex sounds of varying pitch, duration and intensity, such as the songs of birds, are far more difficult to render. In most cases it is even beyond the capability of musical notation to transcribe them in all their nuances.

The problem is compounded by human differences and limitations, a trivial example being the crowing of the rooster, approximately the same the world over. A Frenchman mimics it as "cocorico," but in Germany roosters are heard to have a Teutonic accent and their crowing is rendered as "kikeriki." Italians hear the cock's crow as "kirikiki," and English and American ears hear "cock-a-doodle-do"—or, as Shakespeare transcribed it, "cock-a-diddle-dow." All the while, on the other side of the globe, the Japanese are hearing "kokke-kokko." There is as yet no universal tablature which would enable us to transcribe even this relatively simple call with consistency.

The only scientific solution would be a graph indicating the frequency and duration of the sound, but only a specialist could interpret it and even this method has its limitations, for the birds themselves often add personal variations to the cries and songs typical of their species. They may sing the same song day after day with almost no variation, and then suddenly modify the song within certain limits. The blue jay is capable of imitating with disconcerting accuracy such random noises as the sound of a running faucet, a meowing cat or a squeaking door. Mockingbirds, true to their name, have a large repertory of songs of other species. Almost any bird can and does make subtle modifications in its characteristic song.

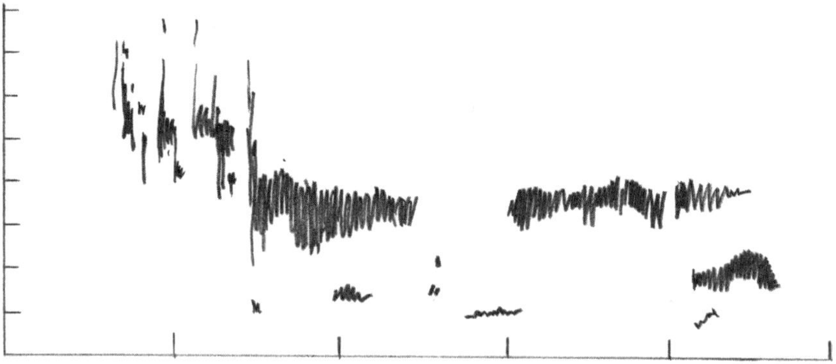

Upper graph: the song of the robin.
Lower graph: the song of the nightingale.

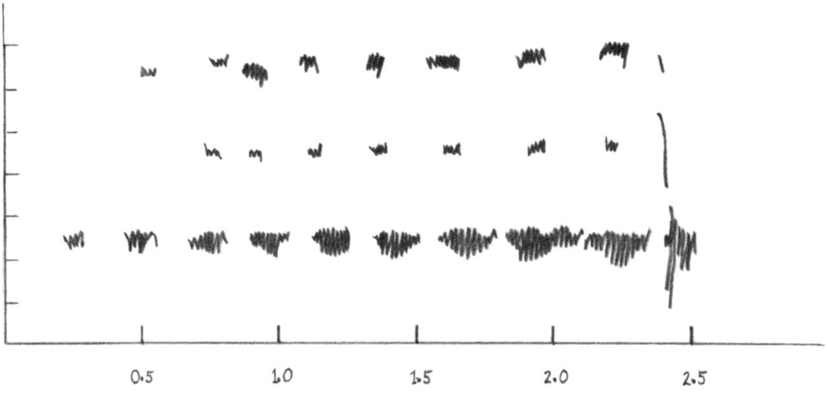

Animals are not as restricted to rigid behavior patterns as we sometimes think. They are capable, within certain limits, of an animal inventiveness, of using their intelligence in a variety of ways within different sets of circumstances.

In folklore and mythology, there are stories of gifted people who could attract animals by singing or playing a musical instrument. The Greeks told of Orpheus, and the town of Hamlin had its Pied Piper. Christianity gives us the legend of Saint Francis of Assisi, whose sermons were so highly appreciated by birds. Even today, such legends persist—snake charmers, alligator wrestlers and the like. In *The Seal and the Girl,* Rowena Farre describes an old Eskimo trapper she has met, a man who attracted wild animals by his mere presence. Nor was he unique. Ms. Farre herself had lived with a relative who also had this mysterious gift, and she witnessed birds spontaneously perching on the woman's shoulders. Obviously this phenomenon is rare if it exists at all, but even without such exceptional gifts, humans can attract many animals by imitating their cries and songs. Hunters are adept at this, often relying on instruments known as "calls" or "lures."

In earlier days, poachers and game wardens made their own simple lures out of cork, bone, shell or shellfish, and even straw, and used them to supplement their extensive knowledge of an animal's habits. In fact, for many of them, mechanical lures were actually unnecessary. Simply by using their hands or voices, they could imitate perfectly the sound of an animal. These people lived in constant contact with nature and were sensitive to the slightest variation in natural sounds. Our modern hearing, attuned to radio and television, is certainly much less acute. But even the most highly trained human ear can hear only a small portion of the acoustic signals emitted by animals. At best, our ears can register sounds which cause vibrations at the rate of twenty thousand cycles per second. Sounds with lower frequencies (infrasound) or with higher frequencies (ultrasound) escape us completely. By contrast, a dog can for example hear sounds of up to thirty-five thousand vibrations per second. Inasmuch as our range is not very large, lures outside the range are of great assistance.

Lures, developed to attract specific animals, promised to make up for these human deficiencies. Most lures were small whistles, in a variety of shapes and sizes, which imitated the sounds made by the animals. Calls and lures are still available, but they have never been altogether satisfactory to the would-be naturalist. For one thing, their mechanical, often metallic sound makes an animal uneasy and suspicious, and it is not so easily taken in. Also, commercially manufactured lures do not always imitate the precise cry that will best attract the animal. Some may actually reproduce the call to flight when it would be much more effective if it gave the summoning cry of females or of the young of the species—or, better still, the distress cry of a predatory animal's usual food species.

For a lure to attract an animal with any success, it must reproduce the exact cry of that species (or of its prey in the case of predators). Even then, it is not foolproof. Living creatures do not always repeat themselves in exactly the same tone. As stated earlier, their language is infinitely more varied than we may think, and emotions such as fear, boredom, pleasure and so on serve to modify it. The cries, noises and songs emitted by animals can be broken down into a series of variables: register (high or low), intensity (strong or weak), duration (short or long) and frequency of repetition (in a given span of time).

As in human speech or song, modulations are important. The voice may rise or descend, tremble, soften, become louder and so on. The variations make commercial lures, and even accurate tape recordings, ineffective in many situations.

Dialects

Ornithologists were surprised to discover not long ago that local dialects exist among birds of the same species. At airports, birds—especially large flocks of gulls—can be a hazard. They are sometimes sucked into the turbines of a jet plane as it lands or takes off, not only killing the animal but damaging the aircraft and possibly causing an accident. In an effort to remedy this in California, ornithologists recorded the distress cries of the seagulls inhabiting the runways and played them back over a loudspeaker, hoping the birds would flee. It was found that, unfortunately, this simple solution does not always work. The experimentors discovered elsewhere that a gull cry can only frighten another gull if they speak the same gull language. The cries of distress recorded in California had no effect on the gulls of Guam which were disturbing the peace at an American airbase. The accents differ.

Elsewhere, this program has been successful. In France, which is so much smaller than the American Pacific Coast, the accents may not vary as widely among birds of the same species and certain birds were repelled successfully from the trouble spots. A team of scientists ridded the island of Dumet and the city of Montluçon of thousands of starlings which were roosting there, destroying the trees and annoying the inhabitants. A recording of their distress cries played through the loudspeaker made the birds fly away permanently (although they only relocated, creating the same problem in the new locale). Though imperfect, this solution is preferable to extermination, at least in the case of those species whose population level does not exceed balanced limits.

Success in chasing away birds by this method is far from guaranteed for other reasons. It did work for a while at airbases in the United States. Then, as time passed, the birds began to adapt to the broadcasts of their own distress cries and ultimately came to ignore them. This same phenomenon has been reported among mammals. For example, the wolf quickly begins to distinguish between the real howls of its brothers and those which have been recorded.

Obviously, the living voice has a quality which is absent in the most scientifically recorded voice, and animals can tell the difference. In learning to identify the cries and songs of animals, recordings and tapes may be invaluable study aids, but they must be supplemented by serious study in the field where natural variations are found. There is no substitute for personal experience and a trained ear.

The Badger

The badger is a member of the weasel family. It resembles a small bear with its thick gray fur and short ears, its flat-footed and rather heavy tread, nonretractable claws and formidably sharp teeth. Essentially a shy and retiring animal, the badger, when cornered, is capable of a strong defense and makes good use of its natural weapons. Like other members of the weasel family, it lives in a burrow about which it is fastidious. Just outside the entrance it digs small holes for defecation and here it scent-marks whenever it leaves the burrow. The badger is as careful with its fur as with its home and scrapes and cleans its coat often.

The Language of Badgers

Because badgers are shy creatures, they are not easy to locate, but observation is made simpler by the animal's extremely regular habits. It leaves its burrow at nightfall, at the same time each day and using the same paths. Every few steps it grunts and sniffs, making it easy to follow. In fact, it is often so noisy that observers can camouflage their own sounds by synchronizing them to those of the badger.

Extremely nearsighted, it depends on its keen sense of smell when out hunting for food, so one must stay downwind. If a badger catches the slightest whiff of a human or other dangerous animal, it will quickly disappear into its

Walking posture of the badger.

burrow. But by standing still and downwind, one can come very close to the badger, which will often accept the observer as part of the natural scenery.

Young badgers play together, yelping like puppies. Their cries—a repeated "pit" or "crrr"—can be heard as far away as a hundred yards. The parents use a short muffled neigh—a "roc, roc" sound—to warn their young of danger. There is one other call which has sometimes, though rarely, been heard: a loud, sharp and terrifying howl whose meaning is not yet known. It is thought to be the cry of males fighting, but an English observer has claimed to have heard such howling near a dead badger and has interpreted it as a mourning cry.

Our Relations with Badgers

It is not quite clear why, but the badger has gained an undeservedly bad reputation for aggressiveness and as a nuisance animal. Perhaps this is because of their ferocity in their own defense, although under normal conditions they will flee rather than fight. Or perhaps it is because they sometimes share their burrows with rabies-carrying foxes and the bad name (though not the disease) has rubbed off on them. In any event, they have been ruthlessly exterminated in many parts of the world. In the past, terriers were trained to hunt them, going underground to chase them from their burrows. The badger had at least some chance against the terrier, but against more modern methods—which include injecting poison gas into the burrow—it is doomed, even when the real target may have been a diseased fox. The fox is usually long gone by way of another exit before the poison is introduced, leaving the slower badger to die in its place.

Whatever the reason, the badger has been grossly misunderstood. A carnivore, it leaves the cabbage patch alone, and instead chooses wasps, bugs, mice and even snakes as food. It is a natural check on the populations of several species which man is constantly trying to control by artificial means. The badger actually does us a great service if left to its own devices.

The Bat

Through no fault of their own, bats have earned a poor reputation among men. They are rumored to attack people without provocation, entangling themselves in their hair. This popular myth may have originated with the bat's hunting technique, which is to emerge at dusk and chase flying insects, sometimes swooping toward the ground. But perhaps the most damage was done to their image by Bram Stoker in his classic horror story, *Dracula,* in which the bat becomes the embodiment of evil—the blood-sucking vampire. Actually Stoker's tale had some basis in fact, for there is one variety of bat which does suck the blood of cattle and other warm-blooded mammals, though rarely man.

Most bats, in fact, avoid man only too readily. During the day they sleep, and during the winter they live in partial hibernation in old trees, abandoned buildings, quiet caves—wherever they can find dark and undisturbed shelter. However, with man's ever expanding population, these shelters are becoming increasingly scarce in many parts of the world. The bat, like other creatures, cannot compete with man and is gradually losing its habitat. In some places it is in danger of extinction altogether.

Bats often hibernate in large groups, and if disturbed they will panic—urinating, uttering high-pitched cries of alarm and flying about haphazardly while trying to locate the source of danger. This traumatic awakening is exhausting and uses precious reserves of energy which the bat needs to survive hibernation. If there are several disturbances during the weeks of hibernation, chances are slender that the bat will survive. As it weakens, its cry of alarm at each new disturbance will become less and less energetic.

The cry of alarm can be heard by most people, as can the "zik, zik, zik" cry of newborn bats. However, most humans whose hearing is average will never hear the faint metallic grating cries of the adult bat in normal communication. Nor can we hear the ultrasonic vibrations the bat uses to navigate during flight.

The bat is a truly remarkable creature on a number of counts. Not only is it the only known flying mammal (flying squirrels do not actually fly, but use a flap of skin to glide from a higher location to a lower one), but it also has an unusual navigation system known as "echolocation"—a term coined by the American scientist, Dr. Donald Griffen, who pioneered this field of research.

Like radar and sonar, echolocation is based on the principle of locating objects by their echo. The bat's "system" consists of a bony larynx emitting ultrasonic vibrations of thirty thousand to seventy thousand cycles per second—high above the range audible to the human ear. These ultrasonic vibrations bounce off obstacles and send back an extremely precise echo which the bat picks up on its "receivers"—highly developed, large, sensitive ears. From these echo returns it can form an accurate picture of the size, shape and even density of an obstacle or of prey, as well as its distance and the direction in which it is moving in the case of prey. The signal is not continuous, but broken by intervals of silence which allow the bat to receive the echo before making another broadcast. It sends ten signals per second at rest, thirty in flight and fifty when approaching an object.

We still have much to learn about bats and their communication. Research has indicated that they use their ultrasonic ranging system to communicate with each other at

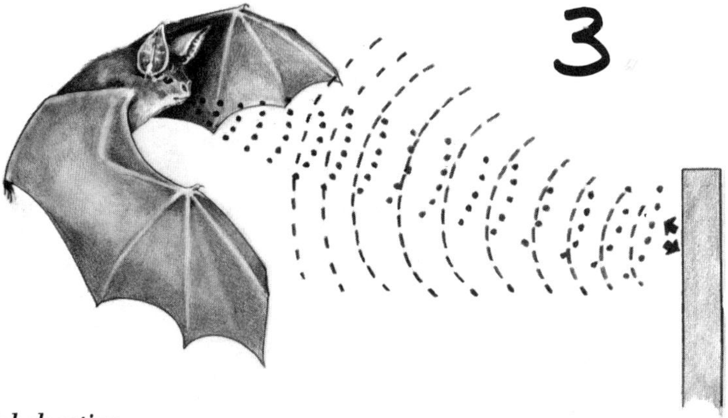

Echolocation.

swarming time and to avoid collisions. Bats may also be sensitive to electromagnetic waves, and it appears that they have a spatial memory map which also aids them in navigation. It has been determined that bats in captivity always recognize the voice of their keeper, the one who feeds them.

Bees

Bees are fascinating and hardworking social insects which have labored in close association with man for thousands of years. Napoleon chose the bee, symbol of industry and order, to adorn his heraldic emblem, which later inspired Victor Hugo to write:

Bees, daughters of light,
Flee from this imperial mantle!

The behavior of bees includes complex methods of communication essential to the functioning of the hive. Through these the labor within the hive is coordinated to be smoothly performed by the different types of bees: the queen who ensures the future of the hive by laying eggs, the drones who mate with the queen and do little else, and the workers, nonreproductive females who perform a multitude of tasks. Workers keep the hive clean and ventilated, warm in winter and cool in summer. They forage for food, gather and store it and scout for new hive sites. All of this activity requires cooperation and communication.

Take for example the forager which has located a rich source of food some distance from the hive. This information must be transmitted to other workers so that the food can be gathered. Rather than simply leading its colleagues back to the source, the forager uses a form of language which Karl

Honeybee gathering pollen to be sampled by bees in the hive.

von Frisch, who first interpreted the phenomenon, called *Tanzsprache* (literally "dance speech") in his definitive study of the honeybee. Beekeepers had observed this waggling dance for hundreds of years without understanding its meaning. The rhythm and choreography of the dance indicate the *distance* of the food source from the hive. A slow dance in the form of a flattened figure eight, or two semicircles joined by a straight diameter line, means great distance. A fast dance during which the forager turns round upon itself in a simple circle means that the source is nearby. The circles are of varying sizes, also relative to distance, but each —even the fast dance—includes a portion over which the bee waggles its body along a straight line. The speed and duration of this waggling indicate distance; the straight line relates to direction.

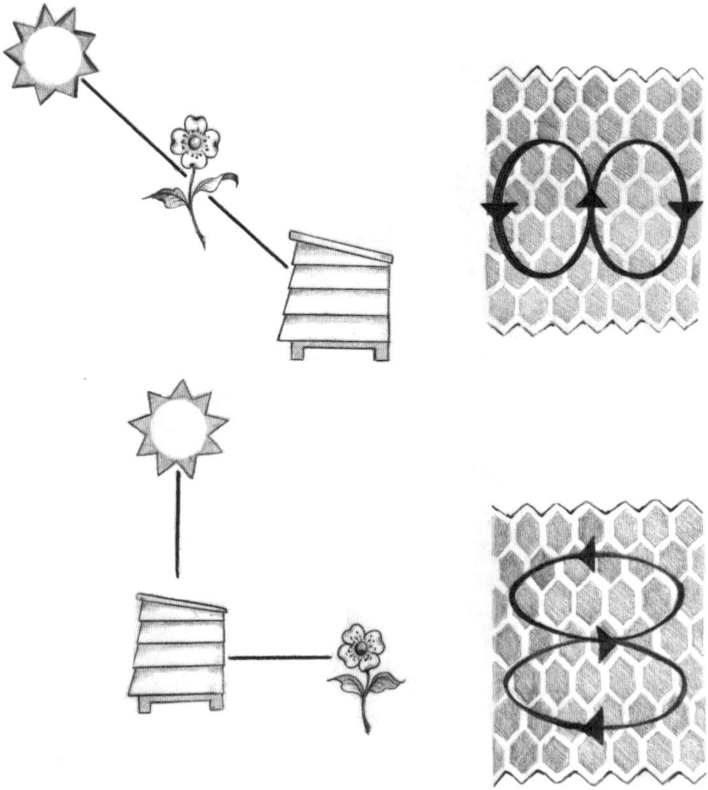

In these two examples of waggle dance patterns, the diagrams on the right trace the movement of the returning bee. In the upper figure, the forager indicates that other bees will find food by flying directly toward the sun; in the lower figure, they will seek out food at a 90° angle to the sun's position.

The *direction* to a food source is described in relation to the sun. Within the hive, it is understood that the position of the sun is always indicated by a straight line, directly vertical. The bee orients its dance to this vertical line by tracing another straight line at an angle to vertical. In this direction the food will be found. For example, if the bee dances straight up, the food is found by flying directly toward the sun; straight down, and it will be found away from the sun; and for intermediate angles, it will be found in the indicated direction. The same method of communication is used to convey information about the site of a new location for the hive during swarming time.

The returning forager has another piece of information to convey: the type of food. It does this by bringing back a sample which the others smell, taste and feel.

The sample serves another function. The dance is not necessarily transmitted visually unless the bee performs it at the very entrance to the hive. Darkness prevails inside the hive and, while still not clearly understood, it seems that other bees use their tactile senses—their antennae—to read the dance of the forager. Their sense of smell alerts them to its presence among them when the forager returns bearing some sample nectar.

Once the basic code is understood, it is possible for human observers viewing the bees through transparent hive sections to interpret the dance and locate the sources of food described by the foragers.

But can man communicate with the bees? Von Frisch, in one of his many experiments, inserted a small piece of iron into a dead bee and, using a magnet, made the lifeless body imitate the waggle dance of the forager. As expected, other foragers read the distance and direction given by Von Frisch and flew to the location he had selected. Curiously, most apiarists do seem to be understood to some extent by their charges. In talking to the bees and mingling with them, the keepers become recognizable and familiar to the bees who will tolerate their presence without attacking, as they would a stranger.

The Blackbird

The blackbird is a lyric artist and never misses an occasion to express itself. A town dweller since the earliest times when cities began to expand, its song may be heard just before dawn in most of the parks and gardens of North America and Europe.

Blackbird Language

Blackbirds employ fourteen discrete cries. Their song, which is extremely musical, consists of a series of melodious, whistled phrases punctuated by a somewhat harsh chirping. This song has a precise significance when sung by males: It indicates taking possession of a territory and serves as the proprietor's "identity card." There is an extraordinary number of individual variations, all of them nevertheless recognizable as the distinctive song of this ubiquitous bird.

Blackbirds imitate each other, and a newcomer to the flock will soon adopt the melodic phrases of its neighbors. This imitation of song is a necessary apprenticeship to their social life; a blackbird raised in isolation will never sing perfectly.

Starting in February, the songs begin in the morning before sunrise. This schedule continues until the middle of July. The most active period of this "bel canto" is the month of March.

When it takes flight upon being disturbed in its bush, a blackbird's cry of alarm sounds like "tsie tsie." Such a rapid succession of syllables is a common characteristic of many species when danger threatens and there is no time for lengthy vocal messages. Less well known is the cry emitted when the blackbird spies a nocturnal bird of prey. This cry alerts the others and provokes them to aggressive action, such as a mass attack upon owls. It can be transcribed as two syllables: "deuk-tik," the first a low note, the second high. Aggressive inclination is also expressed by "tsie," but sung on a note that has a higher tone and shorter duration than does the cry of alarm.

The defense of territory is often accompanied by a cry that resembles the sound of "chink." When the blackbird is on the alert, but has not yet identified a specific threat, it cries "tchor-tchor."

Like many other animals, singing birds have a rich language of gestures. The male blackbird shows its aggressiveness preparatory to fighting by perching forward, its beak open, its body more or less erect. It ruffles its body feathers to augment its apparent size but keeps the head feathers flat. Generally, the tail is dropping and the wings are slightly spread.

On the other hand, a blackbird which has no desire to fight evades confrontations by adopting a submissive stance —beak closed (at the horizontal or below it), tail erect, feathers smoothed back against the body and the wings folded tight. The feathers on its head may be raised, but the overall appearance is one of small size and retiring demeanor.

Postures of aggression are not always easy to distinguish from courting behavior. The male often adopts a courting posture which resembles that of aggressiveness, except that the head feathers are raised. These postures, whether they betoken an aggressive or a courting tendency, are most readily observed during the month of March, the season when courting is in full swing, and the bird makes a special defense of its territory against spring squatters. This aggressive behavior can be found again later in the year, in November, when there is a brief rebirth of courting. In principle, the male occupies his territory throughout the year.

Courting behavior.

When she sees a male arriving to court her, the female adopts the posture of submission but keeps her beak open. The responses between blackbirds are often as ambivalent as they are between human beings, attraction and aversion existing side by side.

Figs. 1 and 2: Submissive postures.
Fig 3: Aggressive posture.

The Black-capped Chicadee

The name accurately describes this familiar little bird. Not only is the crown of its head black, but also its throat. In North America its range is from Alaska and Newfoundland, south to New Jersey, Maryland and Missouri, and during the winter to central Texas as well, although it was long thought that the bird seen in Texas was a different species altogether. During the winter, the chickadee forms the nucleus of heterogeneous flocks of small birds—nuthatches, tufted titmice, woodpeckers, creepers and others.

Conversation

European members of this family, *Paridae,* include several species of titmice, and all of these have several calls and different songs. But the chickadee of North America, though often singing, has only three frequently heard cries. Its song "chicka-dee-dee-dee" is familiar in the north. In addition to an abbreviated version consisting simply of "dee-dee-dee," it makes a clear-whistled "fee-bee" which closely resembles the song of the phoebe.

Body Language

Though perhaps rather limited vocally, the chickadee has an interesting variety of visual signals. One of their preferred foods is suet, and placing a piece on a windowsill in the winter will usually attract several chickadees. The stronger birds will lay claim to the suet by adopting a threatening attitude. They place themselves in a horizontal posture, head stretched forward, to frighten away the more subordinate birds who will withdraw some distance to await their turn.

Menacing posture of the chickadee.

One of the more unusual behavioral traits was described by the American naturalist John Burroughs, who observed a chickadee hen sitting on her nest in the hollow of a tree. He had several visitors who were interested in seeing the chickadee and after a few of them had peered into the hollow, says Mr. Burroughs, "I heard a faint explosion at the bottom of the cavity. The trick of this bird on such occasions is apparently to draw in its breath until its form perceptibly swells and then give forth a quick explosive sound like an escaping jet of steam. One involuntarily closes his eyes and jerks back his head." A harmless but effective way of keeping intruders away from the nest.

Like several other bird families, the chickadee and other *paridae* will readily feed upside down, searching the bark of trees for larvae and insects, although unlike woodpeckers and nuthatches the chickadee does not so easily climb up or down this way.

Chickadees and Humans

Chickadees are very easily tamed and coaxed to eat from one's hand. One way to do this is to dress a scarecrow with the hat and clothes which can later be worn by the one who will feed the birds. For several days sprinkle birdseed on the hat and shoulders to attract the birds. Very soon they will overcome their fear of the human shape and will feed happily. By then donning the hat and clothes and trading places with the scarecrows, one can be very quickly accepted by the birds as a feeding station.

Chickadees will also nest readily in man-made birdhouses provided the openings are thirty-two millimeters wide. This too makes observation simple.

A subordinate bird waiting its turn at the feeding station.

The Wild Boar

By its Latin name, the wild boar is known as *Porcus singularis*—the solitary pig—and indeed the males do go off on their own once they mature. The wild pig is still a close kin to our domestic pig and will interbreed readily, producing fertile young, unlike the nonreproductive offspring of horse and donkey, the mule. When a domestic pig is allowed to run feral, within just a few generations its progeny will resemble its wild cousin both in appearance and habits.

Even without generations of crossbreeding, the wild and domestic pigs share a similar language and many similar patterns of behavior. Their main differences stem from environment rather than heredity. In the crowded pigpens of the modern farm, males must fight continually to maintain their places in the social hierarchy. Freed from the necessity of foraging and the pressures of predation, the domestic pig has the time and energy to carry on these battles for position. His wild counterpart, on the other hand, must concentrate on finding food and evading predators, especially hunters.

Overcrowding has other adverse effects on the behavior of the domestic pig. Both males and females become abnormally aggressive and savage; seemingly unprovoked fighting is relatively common. Fighting between sows in the wild is extremely rare. Actual violent combats between males occur rarely, but they are usually confined to the mating season and stopped before any real damage is done.

Once the male wild boar has matured, at about two years of age, it will leave its troop, rejoining only to mate. The troop is governed under a matriarchal hierarchy, led by the dominant old sow who is most cunning and experienced in the business of survival. Beneath her are several other sows and their offspring. The dominant sow guides the troop through the forest, deciding where they will stop, rest and feed. Attuned to all the tricks of her foes and exceedingly circumspect, she is the first to recognize danger and to alert others to it. She rules for life; the hierarchy is always respected and her authority never disputed.

Wild boar.

Individual members of the troop establish firm ties of friendship. This has been observed in captivity recently, in a game park where two male boars became inseparable. When one of the pair died, the other remained with its body, groaning and seemingly inconsolable.

Language

Vocalizations of wild boars are at a lower pitch than domestic pigs, but their cries and signals are otherwise comparable. Like the offspring of other prey species, the young are almost always silent except when they are threatened, hungry or have strayed. Then they will grunt and squeal in alarm, making a high-pitched and often earsplitting "uie-uie-uie" or "uiek-uiek" sound. When reunited with their mothers, they grunt a contented-sounding "o-o-o." As they mature, their voices become deeper and less shrill.

When the entire troop is threatened by a nearby danger, the dominant sow gives a *cry of alert*—"off-off-off" and the others will disperse, but only to a short distance. If the danger is extreme she will give a real *cry of alarm,* a loud "rouff" or "wouff" which warns the others to scatter in earnest and regroup elsewhere once the danger has passed. For all their noisy foraging when relaxed, boars are capable of a swift, silent retreat through the same underbrush they may have crashed about in earlier.

Boars constantly monitor the air for scents of intruders. They have a keen sense of smell and lift their snouts frequently to sniff at the wind, making a sort of whistling sound. Sows guarding their young will give several "wouffs" to warn of danger, and in response their offspring will scatter and flatten themselves in the underbrush in a maneuver called "leveling."

The angry sow meanwhile will turn to face the intruder and will stubbornly defend her young and her territory against it. She can be a savage adversary and if someone is in her path, she will give a deep menacing warning—a sort of "wouhouh" sound—just before she charges. It is wise to retreat quickly before she moves to attack; many dogs and hunters have been trampled and seriously wounded by a charging sow.

The male boar, when threatened, makes a different sound, known as "nutcracking." He slaps his jaws together, rubbing the tusks in his lower jaw against the molars in his upper jaw. This sharpens the tusks, but it is also a threatening gesture to warn a foe that he means business. After giving this signal, he will frequently charge.

Wild boars use both their keen sense of smell and their own strong scent to communicate, as does the domestic pig. Odor helps a sow locate a strayed piglet and enables the stray to find and rejoin the troop on its own. Individual odors identify members of the troop, and the lingering scent on the leaves and underbrush allows the others to know which animal has passed. It is extremely important in courtship as well, in attracting compatible mates.

Though wary by nature, the wild boar is readily tamed if treated with kindness. A particularly easy way to befriend it is to scatter handfuls of grain in its direction. Very quickly even an untamed boar will come to feed within a meter or two, if not actually eat out of the feeder's hand. One boar was tamed in this way by some members of the Alpine infantry in France. It came to join their parades, marching ahead of the band, and it stood guard outside their barracks. In other cases, baby boars raised inside a home actually come to behave much like pet dogs, following their master from room to room. In situations like these the boar has proved to be intelligent, inventive and entertaining, although as they reach maturity their sheer size and weight can make them impractical as house pets.

The domestic pig has a far higher intelligence than was once thought. His wild cousin, often called the "black beast," exhibits the same astonishing shrewdness. The boar can readily distinguish between a hunter who has set his rifle down and one who is carrying it, and the boar will show little fear of the former while avoiding the latter. Boars also appear to size up the relative skills of different hunters in a group and to locate lines of fire and possible escape routes. They may crouch in the brush until nightfall, but they also recognize amateur hunters who talk and smoke and do not concentrate on the hunt, and it is past hunters like these that a boar will, at an opportune moment, execute its escape—exercising caution and great audacity at the same time.

The wild boar population has decreased as the great forests have disappeared, and those that remain have barely managed to adapt to civilization. In Europe, they travel through towns and villages at night, crossing large highways with great care. At such times they can be observed without fear of attack, providing the viewer remains quiet and makes no sudden or threatening gesture which could frighten and provoke the animal. Most boars will flee rather than fight unless they are badly frightened.

Bovines

A herd of cows peacefully grazing on a hillside may look like nothing more than a number of animals on the loose. But a trained eye will recognize a strict hierarchy at work. Bulls, cows and oxen communicate with each other constantly in a language whose meaning has eluded us for centuries and is only now being deciphered.

Hierarchy

The hierarchy is established by a series of combats or, more precisely, sham fights through which each individual finds its place according to age, strength, weight, sex and so on. Unlike hierarchical systems among other animals, there is no continual struggle to maintain or usurp position. Once the hierarchy of a bovine herd has been established, it is faithfully observed and the animals carry on with the business of grazing and breeding; order reigns on the range and the subordinate animals are submissive to the dominant ones. This kind of hierarchy is efficient, preventing deadly struggles and enhancing chances of the survival of the species.

Dominant animals may be either bulls or cows; sex is not a determining factor. A preliminary challenge, a threatening gesture, is made by one animal to another. The animal approached will either submit to the challenger or make an

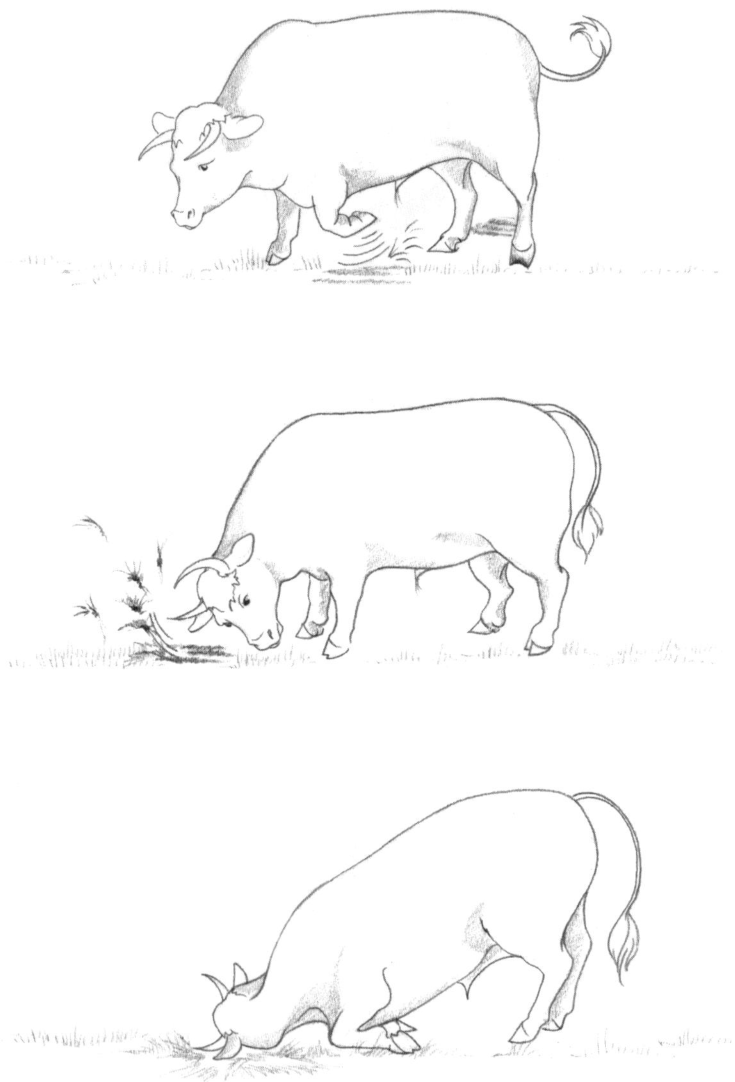

Threatening gestures. Top: Pawing, sniffing the ground and defecating. Middle: The bull digs the ground with its horns when excited. Bottom: It rubs its neck against the ground.

equally threatening response. If the threatened animal is not prompt in submitting or has not detected the threat, it will be butted on the side, neck or shoulder by the challenger. A bull or cow of inferior rank will quickly retreat when jostled in this way.

When giving a *threatening signal,* the animals stand several meters apart. The challenger lowers its head, gazing fixedly at the rival, hind legs drawn in under the body, forehead perpendicular to the ground and horns directed at the adversary. The challenger emphasizes its threat by scratching the ground with its front hooves or by rubbing its neck and head against the ground. As the tension increases, it starts to plow the earth with its horns, tearing and scattering tufts of grass and soil. The two antagonists then approach each other and assume the combat position, either perpendicular to each other (head of one to the flank of the other) or parallel (head to tail). At this point, the weaker animal often gives up, briefly assuming a *posture of submission* and then fleeing.

However, if the animal being challenged is ambitious and returns threat for threat, a real fight occurs. Between two bulls, the combat can be extremely dangerous and may end only with the death of one of the animals. The adversaries first lock horns, then each tries to gain the perpendicular combat position, which will give it the advantage. Once the stronger animal succeeds in placing its horns against the flank of its weaker opponent, the opponent beats a hasty retreat to avoid being gored. The victor will chase it for a short distance to seal the victory.

If neither animal concedes, the struggle continues until one shows signs of submission or until both are exhausted. Each confrontation consists of one or more "rounds," with pauses between them during which the adversaries reassume the threatening position, pawing the ground with the front legs or stabbing at it with the horns. Fortunately full-blown combats are somewhat rare. Most fights are simple

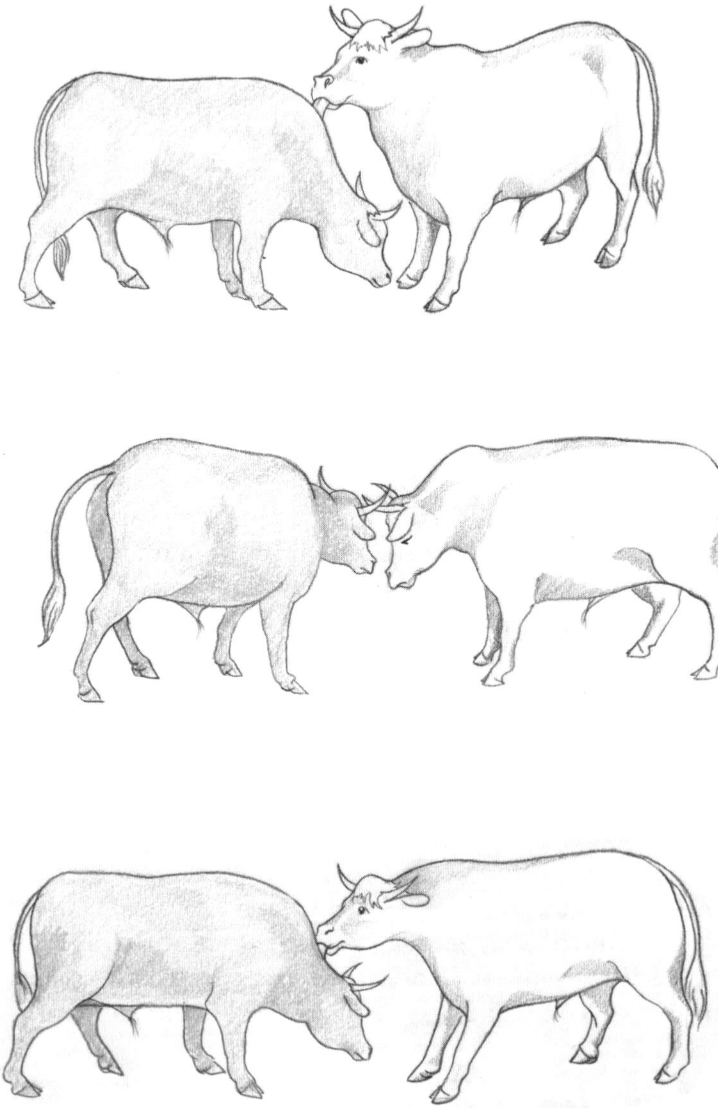

Top: Ox of inferior rank (left) being licked by one of higher rank. Middle: The lower ranked animal (left) challenges, and the higher ranked one lowers and crosses horns in response. Bottom: The lower ranked animal submits and is again licked by the dominant ox, which has withdrawn slightly.

shows of strength—symbolic rather than real—during which the combat postures and gestures are made with reduced intensity and speed.

Among cows, most confrontations are completely symbolic. The adversaries assume a parallel fighting position, head to tail. They slide flank against flank, and neither is really in position to carry through an actual attack. This

The cow in combat. 1. Challenge by the white cow. 2. Head-to-head fighting, each cow attempting to attack the flank of its opponent. 3. Parallel infighting position. 4. Perpendicular position – the white cow has the advantage. 5. The white cow presses its advantage, attacking neck and flank. 6. Mock battle between calf and mother.

position is also assumed by bulls during a long confrontation when neither has managed to place itself head to flank—the stronger position.

Newcomers

Many cattle breeders have learned from experience the importance of the social hierarchy within their herds. Abruptly introducing a strange animal into an established herd can provoke a great deal of disturbance; often the yield of dairy cattle drops significantly. The sudden presence of an outsider has upset the social order of the herd, and the newcomer must find its suitable social rank which means a new series of challenges and submissions. The rest of the herd may exclude the newcomer from the pasture, preventing her from grazing enough to give milk.

Such disturbances can be avoided by placing the newcomer in a neighboring field for a while before allowing it to mingle. It can then get acquainted with the herd and vice versa across the fence. They will observe each other and slowly accept each other. Such introductions are simplest—in fact practically unnecessary—with a young calf which does not present much of a physical threat to the older members of the herd and is thus potentially less disruptive to the existing hierarchy. But whatever the newcomer's age, its situation is at first like that of a schoolboy who arrives at a new school in midterm—the groups of friends have already formed and often unite against the new boy. Many animals behave in the same way, snubbing strangers mercilessly until they become familiar.

Vocal Language

The vocal language of bovines is not yet well understood. We are aware of at least four types of vocal signal given by bulls: the *growl,* which resembles that of a lion and accompanies menacing behavior; a *cry of power,* which usually follows the growl and to which other males respond; the *war cry*, made while engaged in combat; and the *cry of defeat*, given when overcome by a dominant animal. Additional cries, only rarely observed, have yet to be interpreted.

Cows make two different sounds; both are *contact cries*, given by a mother in locating her young. Calves have two corresponding cries to answer their mothers. Variations in the cries of both serve to indicate the distance from mother to calf.

Olfactory Language

Odors play an important role in bovine life. The language of smell permits animals to exchange complex and precise information which is not yet fully understood by researchers. Like their wild counterparts, the bisons, bulls employ their urine as smell "beacons" to mark the limits of their territory. Occasionally a bull defecates while scratching the ground, and although we do not yet know the precise significance of this act, it is clear from observation that it is a deliberate signal.

Breeding

Like most other mammals, cows have a regular estrus cycle, coming into heat several times a year during which periods they can be impregnated. The bull can sense when a cow is a few days away from her fertile period. She gives off a scent which can be detected by the bull's sensitive nose. He will stand beside her, licking her tenderly and sniffing at her with growing excitement waiting for her indications that the favorable moment has arrived. His courtship displays resemble the "threat before combat" behavior, described above. He paws the ground with his hooves and scratches it with his horns, scattering earth, or he rubs his neck against tufts of grass. This threatening aspect of normal courting behavior enables the bull to eliminate or at least

Bull resting his head and neck on cow's rump before mating.

discourage potential rivals. Occasionally he sneezes or sniffs, lowering his head and flattening his nostrils while chasing rivals and dispersing other cows in which he is not interested.

Before the bull actually mounts the cow, he often caresses her with his chin, resting his head and throat on her rump. If she is receptive, she stands still. If not, she breaks away from the embrace. Mating establishes only temporary ties between cows and bulls. There are no stable marriages, with couples united for life. Males and females usually separate shortly after mating.

Although the courting behavior of the female and the male differs, the general behavior of both sexes at other times is similar. Cows also scratch the ground with hooves and horns and rub their necks against it. During combat, they take almost the same stance as bulls do. This similarity of behavior between them has been found to correspond to the great similarity in physical appearance. Conversely, within species exhibiting relatively pronounced sexual dimorphism—where males and females have greatly dissimilar appearances—there are usually more widely differing patterns of behavior.

No discussion of cows, bulls and oxen is complete without mention of the twentieth-century phenomenon, the factory farm. This modern version bears little resemblance to the farms of previous centuries. Animals are confined and crowded in an effort to produce more meat and dairy products for less money. Enormous stresses are placed on the animals, resulting in a wide variety of abnormal and usually destructive behavior and an equally wide variety of illnesses. Ironically, a portion of the money saved by the factory farm system is often spent on antibiotics and other medications to counteract the adverse effects of the system itself. Nutritionists have raised questions about the nutritive value of the meat and dairy products from factory farm animals.

Camels and Dromedaries

Camels are different from dromedaries, yet for centuries the two have been confused. The Asian camel has two humps and frequents the Gobi desert of the Cha-Mo in China and Mongolia where the thermometer can drop to $-40°$ Celsius. The dromedary (sometimes called the Arabian camel) has only one hump, is found in the Middle East and Africa and is the animal which, according to Saint Matthew, can pass more easily through the eye of a needle than a rich man into the Kingdom of Heaven. To add to the confusion, the short hair of the dromedary, prized by weavers and tailors, has long been called "camel's hair."

Both animals have served man as universal draft and saddle animals—the Ford Model-T of nomadic peoples of Africa and Asia since ancient times. They are often called the ships of the desert, for they float gracefully over the sand from oasis to oasis carrying man and his goods. However, indefatigable and long-suffering ferryboats though they may be, they have complex and pronounced personalities and their behavior, once understood, is fascinating.

Language

"Camels use a true vocabulary to express their emotions. Around humans, their tone of voice changes, depending on whether they are expressing joy, fear, sadness, ten-

Camel.

derness or friendliness. This vocabulary includes muffled growls by the male leading the flock; cluckings of defiance and jealousy made to adversaries; love calls during the mating season; cooing murmurs of tenderness by the female calling her offspring; desperate bellows when she is mourning their absence . . . ; drawn-out exclamations of regret when one of the herd is lost or taken away; trumpetings of joy if it returns; contented murmurs by those grazing on appetizing pastureland; bawls of protest by those being loaded down with goods or prodded on to advance."*

Man and Camel

The camel is sensitive to the human voice, especially the singing voice. And camel-drivers have always used an interminable, monotonous chanting to keep the camel from losing its rhythmic pace when it begins to tire during a long trek. A popular legend, illustrating the camel's love of music, records that a certain camel-driver played such sad melodies on his flute that his herd encircled him and wept.

When not chanting, camel-drivers use a universal language in commanding and controlling their animals. These traditional cries have carried over the desert since ancient times:

"Illouk-heylo," a general greeting, friendly and enthusiastic.
"Hab, hab, hab," to summon the camel.
"Woh-ho, woh-ho, woh-ho," to draw the entire herd together.
"Da, da, da," to call back strays.
"Houtch, houtch, houtch," a command to kneel and lie down.
"Biss, biss, biss," the command to advance.
"Wolloo, wolloo, wolloo," the call for return to camp.
"Housh, housh, housh," the command to stop.

*E. H. Finbert, *The Life of the Camel* (Paris, Editions Fayard).

The Cat

The domestic cat, *felis catus,* has shared man's home for about five thousand years, living through times of domestic tranquillity and times of strife. The ancient Egyptian deified the cat, believing that the rays of the sun (also a god) were caught and reflected in its eyes. Their word for cat, *mau,* which one may take to have derived from the cry of the cat, also means "to see," so important did they consider the cat's eyes.

Centuries later, cats had fallen considerably from favor and were associated with witches and devils. In the fifteenth century, Pope Innocent III went so far as to wage an unsuccessful campaign to exterminate them altogether. Their prowess as hunters gradually restored them to their place by the hearth, where they have remained.

Hundreds of volumes have been written about the cat, its history, behavior, care, beauty and so on; here we have concentrated on those aspects of the cat which it uses in communicating with other cats, humans, enemies and prey.

Vital Statistics

EYES. The Egyptians were not alone in recognizing the power of the cat's eyes, as anyone who has experienced a cat's penetrating or menacing stares can verify. There have been many myths associated with the cat's eyes. No less a

1.

researcher than Charles Darwin thought—incorrectly—that "cats with blue eyes are almost always deaf" *(The Origin of Species).* Nor is there any truth to the popular myth that cats can see in total darkness. They *can* dilate their pupils far wider than we can, allowing them to use more of the available light. Alternatively, in very bright light, their pupils close until they are mere slits. Like those of other carnivores, cats' eyes are positioned for optimum depth perception—facing forward—allowing them to judge with fair accuracy the distance to prey and enemies.

EARS. Cats have extremely sensitive ears with a range that extends both above and below the range of the human ear. The cat is capable of distinguishing between closely similar sounds such as that of a family member's car, which it will ignore, and the sound of another car, even of the same make and model, which puts the cat on the alert. Even when asleep, the cat often swivels its ears like radar antennae to train on sounds which are often inaudible to us. They also use their ears to visually communicate in different situations

—pricked forward usually means curiosity, laid flat back for an aggressive posture, laid downward and to the side in submissive gesture and so on.

NOSE. The cat's sense of smell is, like its hearing, acute, though the domestic cat is far less dependent on it than his wild cousins. In the backyard, the cat uses it primarily in feeding, socializing and mating. Like its archenemy, the dog, the cat has a second scent organ inside its mouth. Two small ducts behind the upper incisors lead to this organ. When the cat discovers a particularly intriguing scent, it will lick at it and force its tongue against the ducts to get a better idea of the message the scent conveys. Odors especially help the cat to quickly determine the sex of another cat.

FACE. Some seventy-eight discrete facial expressions have been isolated and described. As all cat owners have observed, the least agitation is reflected in the cat's face.

2.

3.

4.

Fig. 2: Neutral expression. Fig. 3: Beginnings of alert: ears flatten, eyes narrow slightly. Fig. 4: First threatening signs: ears swivel to the back and eyes narrow more; expression becomes set.

5.

Fig. 5: expression of surprise or distrust: ears erect.

6.

Figs. 6 and 6a: Pronounced threat displays.

6a.

Fig. 7: Fixed expression when observing prey.

7.

WHISKERS. These long sensors, called "vibrissae," help the cat to avoid obstacles, especially in the dark. Tip to tip they are approximately as long as the cat is wide, and if the whiskers can pass through an opening with ease so can the cat.

CLAWS AND PAWS. Retractable claws make it possible for the cat to make deliberate statements of displeasure by giving an enemy a good swipe. In anger or fear it will do the same to the back of your hand, and of course when hunting its claws are its primary weapons. But paws and claws are also used to express pleasure. Purring, with eyes closed, the cat kneads on a nubby sweater, extending and retracting its claws in an ecstasy of mock-nursing.

TAIL. When eyeing its prey, the cat often flicks its tail excitedly, but in situations of uncertainty or insecurity the tail is used to signal changes of mood. Two male cats trying to face each other down will switch their tails back and forth, and a cat which is cornered, or frightened (such as a show cat on display), will do the same.

BODY. The cat's body is a marvel of engineering. Its elastic skeleton and loose skin give it remarkable flexibility which, combined with strong muscles in the right places, make it capable of performing astounding leaps. Body language is intrinsic to the behavior of the cat and the message is usually quite clear. It rubs against our legs in pleasure, or it arches its back, spitting and fluffing out its tail to drive away an enemy. Its histrionic rage is a fearsome sight which will frighten off many of the largest dogs.

A more subtle display of annoyance is the cat scratching the ground around its plate. This strange ballet is an elegant way of announcing that it finds its food unacceptable.

8.

Fig. 8: Display of annoyance.

Conversation

The cat is never at a loss for words. It has a wide range of cries and sounds and can change inflections to vary the meaning of even the most ordinary-sounding meow.

The cat's voice, like other aspects of its behavior, has been the subject of misconceptions. In the eighteenth century, the philosopher Dupont de Nemours grandly declared that the cat could pronounce certain vowels and articulate certain consonants. This may have been of some interest to cat fanciers straining to make conversation, but it is really a meaningless observation. The cat is no more likely to need vowels and consonants than we are to speak in meows. The cat has a flexible vocal language of its own that serves it well in its communication with other cats—and indeed with its human friends.

There are over twenty known inflections in its vocal language, and a given sound consistently expresses the same thing.

THE PURR. Physiologically, the purr results from the vibration of the posterior vena cava, one of the large veins which discharge blood into the heart. This vibration occurs when the blood flows rapidly within the chest. The throaty purring of a cat is a sure sign of contentment. Cats purr as they wait for their dinner to reach the table, as they rub luxuriously against a human friend's legs, or as they are stroked and when they are curled in a friendly lap. Ironically, purring is also occasionally the last sound made by a dying cat, as blood makes a final surge to the heart.

There is another sound close to purring, though more like a dry chatter, which is known to many as a death rattle. It is sometimes heard when a cat stalks its prey—particularly inaccessible prey such as a bird sitting on the clothesline or a fly high up on the window.

THE MEOW. This is the basic element of the cat's vocabulary. It can be short and sweet (a simple "hello") or long and drawn out (a mournful plea for food, attention and so on).

A cat will usually make its demands with soft meows at first, gradually escalating to voice its irritation with loud, insistent meows and guttural growls, hisses and finally spitting and sharp cries, almost screeches, similar to those made when its tail is accidentally stepped on.

Meowing, spitting, hissing, purring, wailing, screeching, howling and so on are used to express a cat's pleasure or displeasure depending on the circumstances. Usually there is little doubt about the meaning of the cry. One of the best-known images of the speaking cat is the tomcat on the fence, howling at the moon. This moonlight serenade has been a source of both amusement and irritation to the human audience. This is not the language of love sung by the male to a nearby female cat, but rather a call to cats who are absent.

Pastime, Play and Pleasure

THE HUNT. The cat's prowess in its pursuit of prey is legendary. It appears hypnotized even by the sound of an invisible mouse scratching behind a wall or the sight of a bird out of range. When a mouse or other prey animal is within range, the cat's stalking behavior is a study in concentration and efficiency of motion. Ears erect, eyes open wide, it crouches low to the ground and freezes until theopportune moment when it can best cover the ground between it and its prey. It moves in a slinking run, and if the prey is some distance away it will alternately freeze and run until it is within striking distance. Then, as it closes on its target, it pounces, springing from its powerful hind legs and landing with both front paws firmly on the prey animal's body, pinning it to the ground.

9.

Fig. 9: Stalking position.

PLAY. Kittens and even grown cats are inventive players. They are fascinated by wool, string, bits of paper crumpled and tossed. A toy dragged across the floor becomes an enticing prey creature to be chased. The cat will also swat at a bit of dangling string, or will lie in ambush until an unsuspecting human walks past and will then swat or pounce. Many cats invite humans to join the game. The cat will roll over, switch its tail and wave its paw to say "Come on, let go; play with me."

10.

Fig. 10: Play.

Younger cats play with great concentration over long periods of time. This play is actually practice for the hunt. Older domestic cats quickly grow bored with play—and often with the hunt itself—and pretty soon play turns to sleep.

RELAXATION. House cats, especially older ones, like nothing better than to curl up in a warm, quiet spot for a nap—all the while staying alert for noises that will require investigation. In spite of an easily excitable nature, the cat has a seemingly unlimited capacity for complete relaxation and will adopt the most extraordinary and graceful positions in pursuit of a little shut-eye.

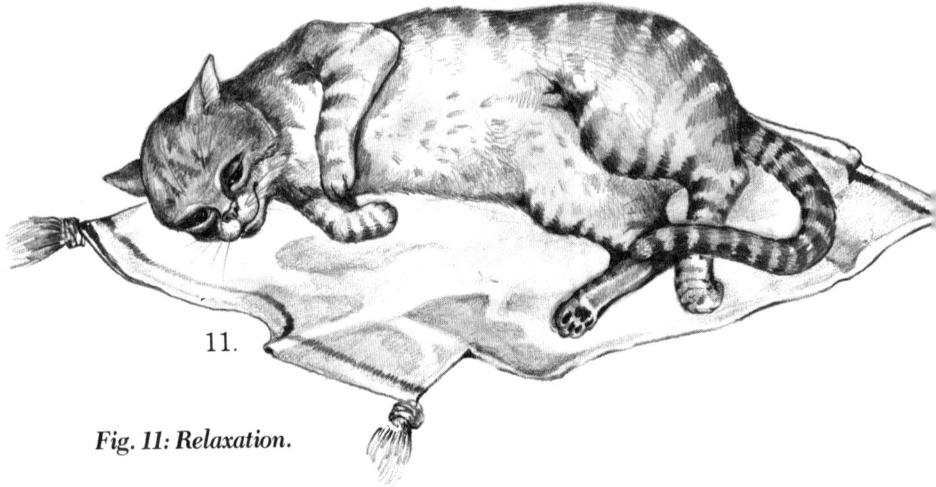

Fig. 11: Relaxation.

AFFECTION. The cat has a fondness for its human host and can be seductive in its displays of affection. It pushes its nose and head under and against the palm of one's hand, demanding a caress. It purrs, rubs against a leg or curls up companionably near. Its responsive displays of affection, observable in the wild, make the cat a rewarding pet in human households.

The Cat in Combat

The cat gives fair warning that it is about to attack. The threatening gestures it uses differ according to its opponent —an importunate human who has gone too far, a dog or another cat being the most frequent adversaries. In most cases, the "threat" is actually a defensive move. The cat is making an attempt to strike a bargain—"You leave me alone and I'll leave you alone."

CAT VS. MAN. When confronting a human being, a cat adopts one of two postures. It may display its fangs and narrow its eyes, or it may keep mouth closed, eyes open and simply look pleading and anxious. In both cases, a front paw remains raised—and it is prepared to strike out, claws extended to scratch. Both postures definitely communicate "stay away."

A cat on exhibit at a cat show is often frightened and will assume a threatening position in its own defense. It flattens itself to the ground, ears also flattened back and the tip of its tail wagging back and forth—a sign of anger. As its hostility increases, it starts to growl softly. With no other place to hide, it huddles against the nearest wall, preparing to strike with its claws at the first hand to approach. This impressive pantomime does not intimidate the judge, however, who has seen it all too many times and is capable of overcoming the cat's fear and overpowering it with firm but gentle handling. For anyone lacking experience in handling a terrified and threatening cat, "hands off" is probably the best policy. If one must handle a frightened cat, one should do so with a blanket which will protect the hands and face.

CAT VS. DOG. The strategy never changes. Standing up with its legs rigid, the cat tries to appear larger than it is. It bristles the hair on its back and fluffs out its tail, holding the tail slightly to the side to emphasize its size and show the dog a formidable silhouette. The ears flatten, the nose wrinkles, the corners of the mouth stretch back and the fangs are well displayed. It growls deeply, ending by hissing and spitting. Often this menacing posture alone will discourage a dog from approaching, or at least slow it down long enough for the cat to escape. But if the cat decides to stand firm instead of fleeing, then the tactic changes. Now the cat arches its back, raises its tail, hair still bristled, ears still flattened, mouth open, fangs still displayed. It is ready to attack if it must. Most of this preparation is simply to make it seem larger than life in a last attempt to scare off the dog. Its posture in fact means that it feels cornered and is terrified and would prefer to retreat the moment the dog's attention wavers.

12.

Fig. 12: Threat display —fur raised.

If escape is impossible, it will give a loud cry of attack and leap like a Chinese demon at the head of the dog, aiming its claws at the eyes and nostrils.

A female cat will use this sequential threatening-and-fighting technique to defend its kittens against all comers. She too stands with back arched and approaches the enemy galloping sideways and swinging her body up and down, so as to present the enemy with a more imposing picture.

CAT VS. CAT. Most cat fights are between two males. The challenger is usually a newcomer who is confronted by the head cat of the neighborhood. The challenger either gives up the fight immediately on confrontation and steals away "sheepishly," as we say, or he will stand and fight.

When the battle begins in earnest, the cats face each other, their legs rigid, without arching their backs or sidestepping. Head to head, they growl, wail and wave their tails. This can go on for some time with all the movements made in slow motion. If neither animal gives up, the two fighters advance, confronting each other side by side, growling, hair bristling, tails beating the air furiously. Each looks for the other's weak spot, and the first one to feel he has found it strikes. The fight itself explodes so quickly that an onlooker is hard pressed to follow the action. The aim of both animals is to use their powerful hind legs and claws to tear at the opponent. The fight ends when the weaker cat has had enough and gives up. Belly dragging along the ground, the loser runs away.

Fig. 13: The loser withdraws.

Territory

The domestic cat's domain is, first and foremost, the place in which it lives—house or apartment. Within that area it will have favorite places—a comfortable chair, radiator, windowsill, basket and so on—to which it will be especially attached. It will defend this territory against intrusion by another cat or a dog.

Outdoors, cats circulate in "corridors," imaginary highways which usually take advantage of the terrain, skirting open spaces, running under bushes and shrubbery and along walls and ditches. A cat considers certain stretches of the corridor—those nearest its home—as its own territory and will mark it as such with urine. An intruding cat following a corridor will stop and retreat when it comes to a recent urine deposit. This "red light" warns that the territory is claimed by another cat. But such marking is only temporary and once the scent is gone, the corridor is up for grabs. This model of a cat's territorial topography is still relatively new but it can be confirmed by observing the behavior of the neighborhood cats.

Chickens

The ancestor of our domestic chicken is the jungle fowl of Australia and Southeast Asia, a ground feeder that rarely flies and can still be found in its native habitat. Its domestication dates from ancient times. It was found in Egyptian poultry yards as long ago as 1200 B.C. Despite this long history of domestication, today's hen and rooster have nearly the same systems of communication and the same social structure as the wild fowl.

Hierarchy and the Rooster's Song

A flock of chickens is organized and governed by a strict hierarchy, in which there are dominant and subordinate members according to age, strength and state of health. An old, authoritative, experienced hen can find herself demoted to a lower rank if she is hurt or sick. A younger hen in her prime will peck her inferiors incessantly to maintain a high rank.

The carriage of the head is an important indicator of rank. A dominant bird holds its head erect, while the subordinate bird carries it tilted. Another indicator of rank is the presence of distinguishing features found on the hen's head. The crest is an obvious and important such feature. For example, if a researcher attaches a false crest to a hen's head, she is treated as a newcomer, loses her rank and falls to the bottom of the social ladder. She must reconquer her place by competing in a series of individual combats. Combs are equally important among roosters.

It is the right of the dominator to peck an inferior without being pecked in return. Dominant birds in the pecking order are also granted first access to the food trough. At feeding time, the dominant rooster summons the flock with a repeated harsh, low clucking. When they approach, only the dominant hen is permitted to eat. With merciless blows of his beak, the rooster keeps away other hens of inferior rank who must eat elsewhere or wait their turn. The dominant rooster has the right to first access to and choice of females with which to mate.

The most familiar cry of the barnyard is probably the rooster's "cock-a-doodle-do." Originally, this was most likely the territorial cry of the jungle fowl, but in the henhouse it has been transformed into the rooster's expression of hierarchical dominance. He crows to maintain position and frequently attacks subordinate roosters if they dare to crow. In fact, many inferior roosters remain virtually mute throughout their lives.

Courtship and Mating

Courtship between hen and rooster depends on an exchange of visual signals—gestures and movements—rather than verbal cries. In general, the rooster takes the initiative in courting the hen, but his behavior differs according to whether he courts a familiar hen or a newcomer.

The rooster performs a simple courtship display for a known bird. He makes his demand by approaching from the rear, head stretched forward, feathers of the neck raised. But when courting a newcomer, the rooster executes a more complex waltz. He turns in a circle around the hen, extending one fan-shaped wing toward her and waving it. At the same time, he takes a few rapid steps interspersed with pauses during which he scratches the ground with one claw.

The hen may respond negatively by withdrawing and making vocal refusals which range from indignant cackling to piercing cries. If her response is positive, she indicates it by hugging the ground, head lowered and wings spread.

Courtship display: The rooster circles the hen.

Communication Between Hen and Chick

Verbal cries play an essential role in the relations between hen and chick, beginning even before the chick has hatched. While still inside the egg, the unborn chick actually learns to recognize the clucking of its mother. To demonstrate this, a recording of the hen's cluck of reassurance was

Top: The hen sees but cannot hear her disturbed chick, and ignores it. Bottom: Now, although she cannot see her chick, she quickly homes in on its distress cry and comes to the rescue.

played near the eggs in an incubator. After hatching, the newborn chicks were placed between two loudspeakers, the first playing the clucking record the chicks had heard, and the second playing the clucking of another hen. Without hesitation, the chicks approached the first loudspeaker as they would have hurried to their real mother. The experiments proved that each hen has a distinct voice and an individual clucking sound and that the chick's recognition of this voice is acquired, not innate. Similarly, the hen's cry of alarm is understood by her chicks, which seek cover in silence when they hear it.

The unhatched chick is not, as one might think, completely passive. In addition to learning to recognize its mother's voice, it makes little cries of displeasure. In response, the brooding hen clucks reassuringly, and the chicks then make cries of contentment. After hatching, the chick's exchange with the hen increases. Cries of displeasure become more intense piping cries, which help the hen locate members of her dispersed brood quickly.

Sight plays no role in the initiation of the hen's search. In another experiment, scientists placed a chick under a glass bell jar so that the hen could see but not hear it. She ignored it completely. Then a chick was placed out of sight but within range so that the hen could hear its piping clearly. Immediately she homed in on the signal and located her disturbed offspring.

The Dog

Dogs have probably been domesticated longer than any other animal, so long in fact that the earliest days of their partnership with man are shrouded in the fog of prehistory. But in spite of man's close association with dog as companion and co-worker, most people—even dog owners—do not usually take the trouble to know and understand the dog fully.

Unlike cats—their only real competition as most favored species—dogs come in a great variety of sizes and shapes. Some have a keener sense of smell than others, some have better vision or more acute hearing, while still others are born to run, chasing over the plains after deer and other game. Most of this specialization is the result of controlled breeding. Different breeds of dogs have been developed to perform certain tasks and so need one sense more than another. This makes it impossible to generalize about canine senses the way we can about a cat's. However, dogs do have certain things in common in terms of behavior, communication and body language.

Conversation with Humans

A moderately bright dog can be taught to understand about fifty human words, ranging from commands such as "sit," "heel," "stay," "come" and so on, to pleasure words such as "go out," "biscuit," "dinner," to specialized terms used among certain working breeds—hunting dogs, sheep dogs and so on. The size of the vocabulary a dog recognizes depends on which words the dog's master chooses to teach it, or at least consistently uses. Even untrained dogs learn almost in spite of themselves, provided they are repeatedly exposed to the same words in the same context, and with the same inflection.

Dogs also respond to a human's *unspoken* language. They are quick to interpret messages we signal with our bodies, even unconsciously. For example, a dog will sense immediately when its human companion plans to go for a walk without it and the dog will look properly dejected. It also senses its human companion's feelings toward a visitor and will adopt the same attitude toward the guest, although sometimes it will make up its own mind in distinguishing between a human who is afraid of it and one who is not. It will sometimes take an aggressive stance toward the former while welcoming the latter.

THE BARK. The dog has a repertory of barks whose meaning is usually clear. A series of sharp warning barks usually refers to imminent encroachment of the dog's home territory and says "be careful; something is happening. Someone is approaching." It may also be used outside the home to tell another dog or a person to stay away.

The dog also has a happy bark used to greet its returning owner. Accompanied by leaps and sidesteps, this bark signals joy and a desire to play.

Furious barks are accompanied by bristling fur, bared fangs, stiff legs, the tail held in a certain position and the face reflecting the dog's mood. It is obvious that the dog is expressing violent anger, a desire to attack or defend itself and its territory—in other words, expressing a threat. However, submissive animals may also behave this way when they are badly frightened. A strong offense may, for them, be the best defense.

Whining is used to express both pleasure and pain. Little whines accompanied by a wagging tail (Figure 1) and leaps of joy are used in greeting and making requests such as asking to be taken for a walk or be played with.

On the other hand, a sad dog, tail between legs and perhaps scratching on a door to be let out, will whine mournfully. It feels abandoned or punished and is crying and begging.

But a dog does not express itself with voice alone any more than any other animal—or human, for that matter. It has a rich and varied language of gestures, postures and expressions. Entire books have been written about the behavior of the dog, some devoted to one specific breed or another, and here we can only suggest the range of behavior used in communication. We have concentrated on the most common behavior and hope that interested observers will want to explore the subject in greater depth elsewhere.

Body Language

The way a dog holds itself and maneuvers its ears, tail, eyes, mouth and the rest of its body delivers a concise message to other dogs and to human observers once the language is understood. Most of the elements of this system of communication are learned, not inherited. Training is usually completed between the third and seventh week of a puppy's life. During this crucial period, the puppy must have some contact with peers, adult dogs and people so that it will be well adjusted when it matures. It also needs varied stimuli in its environment and must not be kept confined in a cage even if in the company of its mother and littermates. The United States Army learned about the importance of a varied environment when it raised and trained pups in a kennel. When they were taken to the woods to put what they had learned into practice, the dogs panicked, cowering and defecating in fear. It took many months for their trainers to overcome this canine agoraphobia which is known as "kennelosis."

The play language of a young pup is a good starting

point for a discussion of canine communication. The messages are clear. The puppy simply bows to a potential playmate, its front legs almost horizontal to the floor, its hindquarters upright. It wags its tail, cocks its ears, and utters excited little yelps or opens its mouth in a wide panting grin. That wagging tail and facial expression are unmistakable. In fact, tail and face are two of the best indicators of the dog's mood.

THE TAIL. The position of the tail may signal play, as in the case of the puppy, but it can also give a warning, especially when combined with other signals. The chart on page 80 shows nine common tail positions. In Figure 1, the tail is lowered and wagging and the dog is friendly and submissive. In Figure 2 the tail is partly raised also meaning that the dog's interest is aroused. It holds it in this position when sniffing at some new and unidentified scent. An upright tail wag as in Figure 3 means that the dog is excited. It can be either an aggressive wag (when accompanied by raised hackles—the hair above a dog's shoulders) or a friendly wag when the dog is investigating another dog, especially when a male is aroused by a bitch. The lowered tail in Figure 4 is the neutral position. The dog is at rest and at ease. In Figure 5 the tail is turned in slightly. The dog may be apprehensive or may be indicating submission. Many dogs hold their tails this way while eating. In Figure 6 the slightly raised tip also shows that the dog is aroused. It is usually searching, exploring and appraising the situation. The oddly rippled tail in Figure 7 indicates tension but not necessarily fear. In Figure 8 the tail is hanging in an intermediate position between Figures 5 and 9 and it indicates that the dog is definitely apprehensive, slightly on the defensive. The sick or completely submissive dog either lets its tail droop straight down as in Figure 9 or tucks its tail in tightly between its hind legs. This latter position also indicates great fear.

There are some dogs which have a very limited range of expression with their tails. Eskimo dogs, for example, and

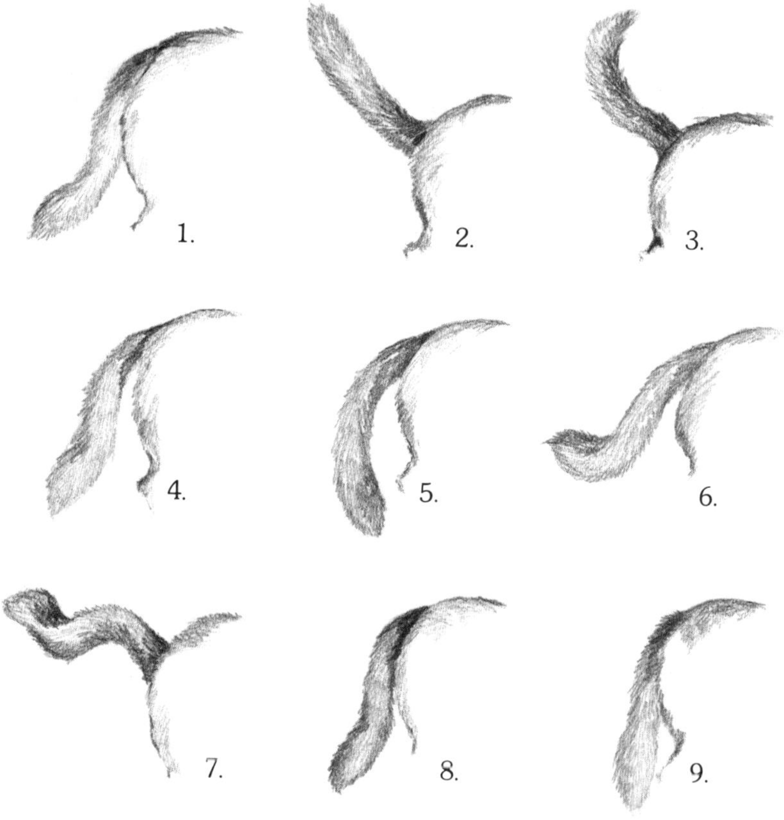

bulldogs with twisted, round tails have lost a great deal of tail communication, as have dogs whose tails are clipped to conform with human ideas of beauty.

THE FACE. Tail messages cannot be interpreted accurately in isolation. The dog's facial expression must also be evaluated to determine the dog's mood. In the neutral position, all the features—including the ears—appear relaxed. From this position, the dog can shift its features to display aggression, submission or a combination of the two.

10.

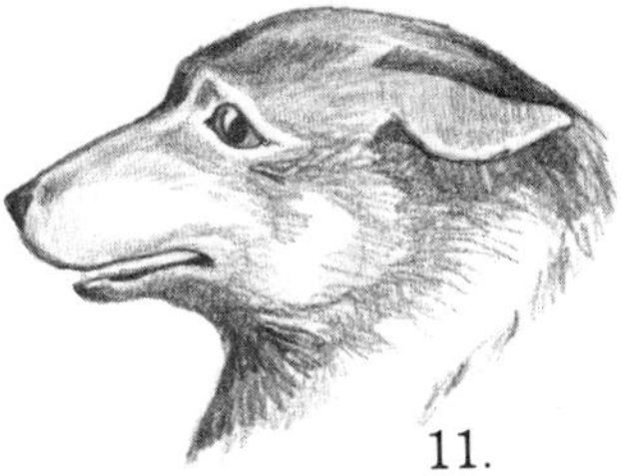

11.

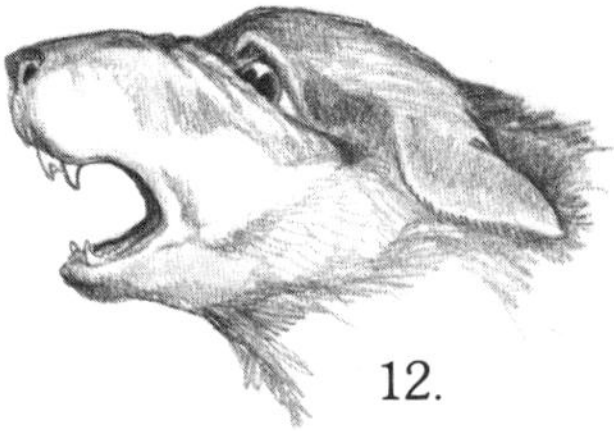

12.

13.

14.

15.

In Figure 10 the dog displays maximum aggression. The ears are cocked forward, the nose wrinkled, mouth partly open and pushed forward, teeth and fangs well exposed. The dog in this posture will be growling, its hackles raised and its tail erect. It means business.

At the other extreme, the dog in Figure 14 is displaying complete submission or fear. Its mouth is closed and stretched back in what is often described as the submissive grin. The ears are flattened and the entire demeanor is one of appeasement. This dog will hold its tail tucked between its hind legs and crouch down low to the floor. The facial expression in Figure 15 is very similar, but the flattened ears are held slightly to the side and the mouth is not fully stretched. This dog is probably ill and its response is quite similar to that of a frightened dog.

The dogs in Figures 11–13 reflect a confused mixture of aggressive and submissive attitudes. The wrinkled nose and snarl in Figure 11, combined with the flattened ears, indicate an aggressive but frightened animal. This type of behavior is common to the type of dog best described as the fear-biter, which, as the name suggests, will bite when it is afraid. Most bites occur when people have misinterpreted the signals as ones of submission.

In Figure 12 the dog displays less aggression and behaves more uncertainly. The eyes are open wider and the fangs are not nearly so prominently displayed. The mouth is stretched submissively although the ears are still raised. In Figure 13 the dog has backed off even more from an aggressive stance and, although it is still menacing, it is not as much of a threat. Its uncertain behavior is leaning more toward submission, and it should be relatively easy to win this hesitant dog over.

Outside the range of aggressive and submissive facial expressions, there are a few others which are frequently observed. First is the wide, open-mouthed, panting grin used to solicit play from other dogs or from humans. It is commonly seen in puppies but is not uncommon in grown dogs. There is a milder version of this expression which is known as the "greeting grin" and is seen in dogs and their wild cousins, wolves and coyotes. It resembles a human smile—lips pulled back, corners tilted slightly upward and teeth sometimes slightly exposed. The ears are usually flattened submissively, but the tail is wagging as the dog moves forward to greet a canine or human friend. Though less frequently seen, there is an even subtler greeting grin shown by some bright dogs who are able to mimic their human hosts. The dog raises its lips vertically to expose its front teeth. In effect, this grin resembles a snarl, but when seen in context of the other body signals—ears flat, wagging tail, playful movements and so on—the invitation becomes clear.

THE BODY. As suggested above, signals of face and tail relay only part of the message. The dog also uses its body to reinforce the meaning. An aggressive dog raises its hackles and hair, and stands tall, legs rigid, leaning slightly forward, tail raised and the appropriate expression of aggressiveness on its face. A submissive dog cowers to the ground, making itself as small as it can and gesturing submissively. (See Submission, below.)

Fig. 16: Relaxed and friendly. Fig. 17: Potential fear-biter.

Most of the body, face and tail signals are used either to warn other dogs and potential foes to keep their distance or to invite dog and human to approach and play. In Figure 16, the dog is relaxed and friendly but alert—ears forward, mouth stretched and open, tongue out. The dog shown in Figure 17 is probably a fear-biter. Its fur is raised, ears back, mouth stretched and open, eyes partly closed, tail partly raised.

Submission

The dog's social system is based on hierarchy. In the wild it would live in a pack similar to those of wolves, coyotes and other wild species. Each dog would find its place beneath one dominant male dog, and in linear fashion each dog would submit to those above it in the order and dominate those beneath it. In the domestic situation, in the home, most dogs will look to their owner as they would the dominant dog in the pack and will be submissive—some animals more so than others. It may also allow itself to be dominated by a household cat or another dog.

A submissive dog will approach a dominant dog and invite attention. It bows slightly, lowering its tail, flattening its ears close to its head. It may flick out its tongue in a licking gesture and touch the dominant dog with its muzzle or, more rarely, raise a front paw against the dominant dog's flank, wagging its tail and alternately raising its front paws. (See Figure 18)

If the dominant dog is receptive, it will respond by investigating the other dog, sniffing at its genito-anal region. During this inspection, the submissive animal will stand perfectly still. In the most passive position, the submissive dog lies on its side and raises a hind leg and the corresponding front leg. Its tail is tucked in tight between its hind legs, its ears are flattened, mouth stretched and it may flick its tongue. (See figure 19)

Fig. 18: Dominant puppy (left) keeping an inferior ranked puppy in its place.

It is this position which a puppy presents to its mother which will respond by licking and cleaning the pup's genitals. In fact, during the puppy's first few days of life the mother completely controls her puppy's eliminative behavior, stimulating elimination by licking. She will eat the feces and lick the urine to keep both puppy and den clean. This behavior can be traced to wild ancestors and is observed today among wolves and other wild species. The mother is actually protecting her litter by confining the puppies' scent to the small den area during their first days of helplessness. In this way she can leave them while she hunts and forages for food.

Domestic dogs which are especially submissive will present this position to their dominant humans, and some will even go so far as to urinate, usually no more than a couple of drops. They should not be punished for this involuntary act, which in fact is a signal of their respect for their owners.

Fig. 19: Position of passive submission.

Another frequently observed gesture of submission is the bowing dog whose throat hugs the ground while it attempts to touch the muzzle of a dominant dog with its own muzzle. The dominant animal may respond by closing its mouth over the supplicant's head and growling. A dog which has been scolded by its owner will behave in a somewhat similar way, crawling forward, belly to the ground, tail between legs, licking and begging for forgiveness.

Courtship

The dog performs a complex ritual of courtship, involving a sort of mating dance by the male. He circles, sniffs, licks, bows to, even embraces his partner before mounting. She may stand perfectly still, indulging in only a little reciprocal licking, or may attempt to flee—sometimes successfully—if she is not receptive to the male's advances. She may also turn and snap at him.

Some dogs do show clear preferences regarding their mates. Males will ignore certain bitches even when they are in heat, and bitches will drive off males they find unappealing. This selectivity is even stronger among wild canines. In the interest of ease in breeding, the instinct has been almost completely bred out of our domestic dogs. Breeders want to be able to improve a breed by choosing which dogs will mate, and so discourage selectivity. If the choice is left to the dog, the breeder might have little chance to change or improve the breed's dominant characteristics.

Territory

In today's cities and suburbs, many of the aggressive displays made by dogs are in defense of their territory. Dogs, like other animals, have a strong need for and sense of territory which they will defend against intrusion. Territoriality is more easily observed in a suburb or rural area where the dog spends a good deal of time outside on its own. Each dog carefully marks the boundaries of its territory with a few strategically placed drops of urine. If it comes to the scent-marks left by an intruding dog, it will immediately cover and thereby cancel the rival's marks with its own urine. A visitor's car will be scent-marked to signify that the dog now considers it part of the property.

Dogs prefer to scent-mark with urine, but they will occasionally defecate as well. Usually, however, excrement is left far from home, along one of their regular routes. Occasionally the dog will follow the act of scent-marking by scraping the ground vigorously with its hind feet as a visual territorial delimitation.

In the city, the dog must share its outside territory with many other dogs, but it spends a great deal of time sniffing at the scents left by other dogs and interpreting the information they contain. The dog will cover some odors instantly with urine and will ignore others. Often it will lick an especially interesting or perhaps unreadable scent and allow its second sense organ inside its mouth to sift through the information.

The dog's highly developed sense of smell enables the male to identify and locate a bitch in heat with great accuracy. Males have dragged their owners off the beaten track to a certain window or building where they know there is a bitch in heat. They are rarely wrong.

The Donkey

Donkeys have served man as beasts of burden far longer than their cousin, the horse. Since biblical times the donkey has been carrying loads to market or pulling carts and even plows. As its Latin name, *Equus asinus,* suggests, the donkey is more accurately called the ass. But there are other names as well: jacks and jennets (or jennies) are large male and female asses, respectively. And the mule is an infertile cross between a jack and a mare (female horse). A less common cross is between a jennie and a male horse (stallion) and is called a hinny.

Conversation

Donkeys are usually rather quiet, except during the breeding season. Then the donkey stallion makes a trumpeting call, a braying well known to all as "hee-haw." After five or six brays, he exhales deeply making a sighing sound like that made by a pair of blacksmith's bellows. The female replies with a softer bray to which the male responds with another series. If there are several donkeys in the area, they can create quite a din.

Donkeys and Humans

The donkey's reputation for stubbornness is actually

Donkey.

based on its natural response to an adverse situation. Unlike the horse, which would panic, the donkey's first instinct is to freeze. When treated with gentleness, it should have little reason to be recalcitrant. Many people find the donkey more intelligent and responsive than the horse, and in fact it can exhibit a doglike affection for human companions.

The Dormouse

The dormouse is a small, nocturnal rodent, native to Europe, Asia and Africa. It has dark rings around its eyes and gives off a very strong odor of musk which permeates its flesh and fur. This scent has helped the dormouse to survive by making the animal unpalatable to predators and repellent to hunters, furriers and laboratory scientists.

A family of dormice will announce its presence in a human home by invading the larder for nuts, seeds and fruit. It will choose the ripest fruit, even going so far as to nibble only on the ripest side of the one selected. Dormice are most frequently evident in old houses and will make their nests, which resemble bird nests, in cracks in the walls. In temperate climates, they hibernate through the winter months curled into tight balls in their nests. If possible they will avoid all contact with their human hosts. They do not make amiable pets and can be tamed only with great difficulty if at all.

Communication

The strong musk scent that the dormouse secretes is its primary channel of communication. The muzzle, head and flanks of a newcomer, and the genitalia in the case of the female, are very carefully investigated by an established group of dormice to determine the sex, identity and intentions of the stranger.

Dormice are territorial and will rub their scent against stones, trees and the ground to scent-mark the boundaries of their territories. The malodorous marking may protect them from predation, discouraging or actually repelling predators.

Grooming.

Other signals are used in communication. Dormouse behavior includes the grooming ceremony. One animal will lick or nip the neck, nape, jowls and top of the head of another. This intimate ritual is usually performed between members of the same sex, living in the same group, although it has been observed between a dominant male and a female. The social structure of the species is based on hierarchy, and grooming activity is considered the privilege of the leader, usually a dominant male. He, or occasionally another high-ranked individual, will initiate grooming. Higher individuals are always groomed by lower ranked animals. Through this behavior, the social ties of the group are reinforced and new members are integrated into an established hierarchy.

Another posture often observed is the "boxing" stance adopted by two animals who are meeting for the first time. They raise themselves up on their hind legs, front legs held out rigidly, while they take stock of each other. Dormice also vocalize to communicate. During aggressive encounters they make at least six different cries, each of which is a reaction to a different situation.

Top: Boxing stance.
Bottom: On the alert.

The Fat Dormouse and the Tiny Dormouse

These two species are cousins of the common dormouse and share the same behavioral characteristics in spite of having different appearances.

The fat dormouse resembles a gray squirrel, although its head is closer to that of a rat. It has a tufted tail and gray fur on its back with white fur on its belly. Its Latin name, *Glis glis,* is a more attractive one but it may owe its English name to the Romans as well, for they raised and fattened this species of dormouse. In spite of the mouse's foul odor, the ancient Romans considered *Glis glis* a great table delicacy.

In fact, centuries later, they were still being eaten. Gustave Flaubert, in his African novel, *Salammbo,* describes a meal of candied dormice in honey, served between two other gourmet delicacies—camel's leg and fried grasshoppers.

The fat dormouse has a wild disposition and will bite at the least provocation, especially if it is disturbed in its nest during hibernation.

The sleepy dormouse Alice met at the tea party in Wonderland was most likely the muscardine, the tiny dormouse. It is an exquisite animal with large black eyes and as its name suggests it is quite small. Unlike its cousins, it can be easily tamed, especially when provided with its favorite foods—raspberries and hazelnuts. It will even tolerate being carried about in a coat pocket and in fact becomes very friendly and domesticated. During hibernation, the muscardine curls up in a ball in its nest and will not awaken easily even when handled, although it may whistle slightly if the handling is rough.

Unlike other dormice, the muscardine avoids human habitations when foraging. Otherwise, its behavior and communication system is similar to the other species.

The muscardine, or tiny dormouse.

Wild and Domestic Ducks

There are almost one hundred fifty species of wild ducks found in America, the most familiar and widespread being the mallard. There are also several species of domestic duck of which the large white Peking duck is probably best known. Like many others, it was bred not to fly, but otherwise the Peking duck shares many traits with the wild mallard. For one thing, the Peking was developed in China from the mallard which had been domesticated to provide a much needed source of protein for that populous country. The Peking and the mallard speak the same language, albeit with different accents, and in general their behavior is much the same. They will interbreed readily.

Conversation

Mallards are the most talkative of the wild ducks. When safe in a secure location, they quack incessantly. It is the loud quack of the female mallard which has come to typify the cry made by ducks everywhere. Males produce softer reedy notes and ducklings make a whistling sound to attract their mother when they become separated. But it is the quack or nasal "reb" sound which is of interest here, for it is the most frequently heard and has a variety of meanings according to its volume and duration.

When faced with danger—a dog, for example—the male gives a general alarm by means of a single loud and prolonged "reb." On the other hand, a repeated "reb-reb-reb" made by a female will mean either that she is changing activity or shifting location. Males too make this cry just before flying to a new locale. A quieter version is made by both sexes as they walk about on land and it is given incessantly by the mother duck as she escorts her ducklings. Apart from the whistled alert already mentioned, ducklings have a special cry used in ordinary conversation—"qua-gae-gae-gae-gae-gaec." The second syllable is stressed by the volume increases from start to finish.

After a long separation, males and females greet each other with a characteristic double "reb-reb." Males, or drakes, make a long, tremulous reedy sound to attract females. Among ducks, the female selects her partner, sometimes refusing a prospective mate, even driving him off with a loud distinct "gek-gek-gek" which sounds very much like coughing.

Body Language

The ritual of courtship includes a series of gestures and ritualized movements.

Courting takes place in the water, usually on a quiet pond. The drake faces the female, bows and then raises his head and neck. The female responds by stretching her body horizontally on the water. Male and female repeat these two basic postures—sometimes for hours—before they actually mate.

All ducks follow similar courtship rituals, although there is some variation among such types as teals or pintails. And nesting habits vary, with some species choosing sites off the ground and others making their nests in grassy marshlands.

Top: Steps in the courtship display of the male. Bottom: Complete sequence of displays. 1. The male shakes his beak; 2. stretches his neck; 3. shakes his tail; 4. bows; 5. raises his tail and wings; 6. looks toward the female. In response the female 7. stretches horizontally on the water's surface and 8. looks toward an encroaching male (opposite page).

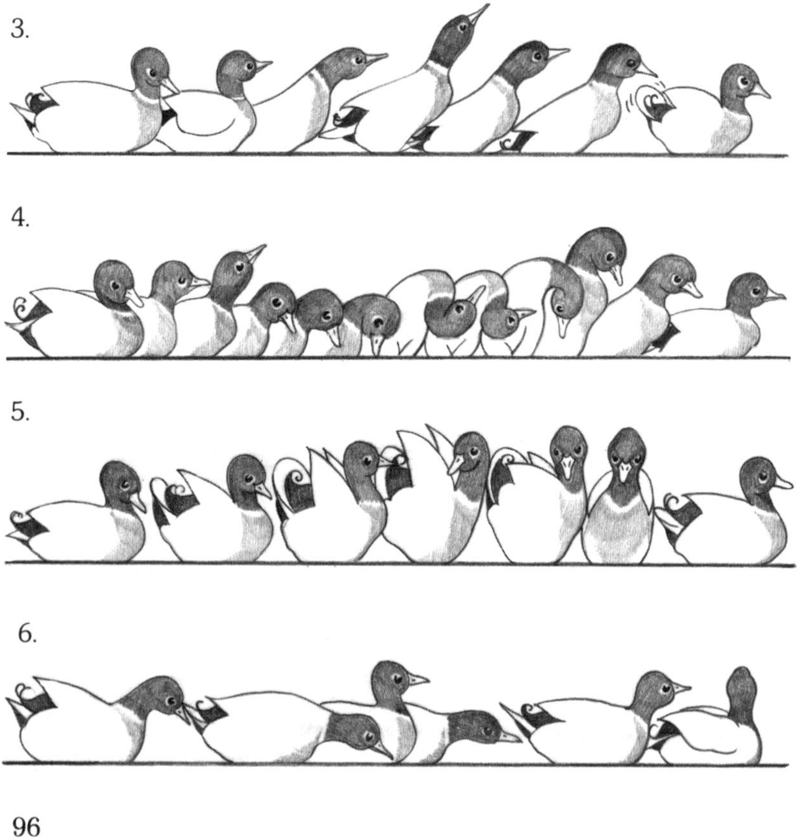

Female (right) greets the male.

Ducks and Humans

Mallards have adapted themselves well to the presence of man. They inhabit his ponds and reservoirs and even feel at home in city parks and waterways. Historically, most of man's contact with these birds has taken the form of predator-to-prey, and man learned to communicate in a rudimentary way simply to make hunting ducks easier. Wooden decoys and duck calls were—and still are—used to attract wild ducks and make them better targets for the hunters waiting behind their blinds. These same devices are sometimes used by bird watchers to get a closer look at their subjects, although their use is sometimes not considered orthodox.

Newly hatched ducklings, like the young of other species, will "imprint" readily. Imprinting behavior is the tendency of young animals to seek and attach themselves to a mother figure. The young duckling emerges from the shell and immediately begins to look around for its mother. If its real mother isn't available, it will attach itself to the next best thing it can find. Ducklings raised with chickens will imprint on a chicken and behave as it does, even refusing to swim. They will also imprint on a human being. Konrad Lorenz studied the imprinting behavior of ducks and geese, concentrating on the latter. (See The Goose.)

The Elephant

While the elephant is essentially a wild animal, it is easily domesticated and its great strength has long been put to the service of man. There are two species—Indian and African—the Indian elephant being slightly smaller and easier to handle. It was the Indian elephant which accompanied Hannibal across the Alps, although accounts still vary as to how or precisely where this remarkable feat was accomplished.

Communication

Elephants have a limited ability to vocalize. When excited or threatened they respond by grunting, and as their excitement mounts they "trumpet," a sound familiar to anyone who has visited a circus or a zoo at feeding time. This cry alone, when the elephant is angered, is enough to discourage man or animal from approaching. It is a loud, prolonged sound sent out through the elephant's trunk and can be heard for miles. Volume and duration will alter the meaning. A hungry elephant in the zoo will give a short, sharp trumpet to demand food, whereas an angry, charging elephant in the wild will give a loud, long, threatening trumpet.

Body Language

The elephant has been the subject of several behavioral

studies, some aimed at enabling governments and game wardens to protect and control the remaining wild herds in Africa. In the process, we have learned much of the body language of the elephant. The position of ears and trunk are important indicators of the mood of the animal. The trunk is to the elephant what the tail is to the dog. An irritated, aggressive elephant will straighten its trunk and hold it quite rigid while raising its ears. On the other hand, a fearful and defensive elephant will curl its trunk downward while lowering its ears submissively. These are clear-cut extremes, but as with other animals there are many intermediate and mixed signals which reflect conflict within the animal—uncertainty whether to stand and fight or to retreat.

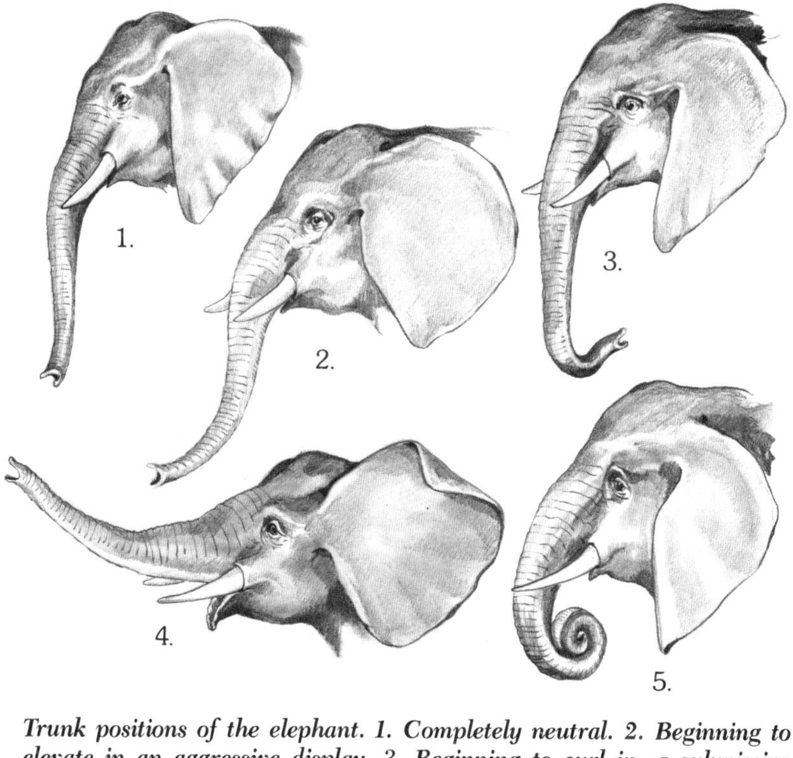

Trunk positions of the elephant. 1. Completely neutral. 2. Beginning to elevate in an aggressive display. 3. Beginning to curl in, a submissive display. 4. Aggression. 5. Submission or fear.

Elephant herds are governed by a hierarchical system with dominant and inferior members. However, unlike other herd and group animals, elephants do not have a true contact cry. Because of their great size, such a cry is simply not necessary. Also they are very noisy as they forage and as they digest their food, and these sounds may take the place of the contact cry. Scent glands along the temples allow individuals to identify each other and permit families to regroup with ease.

Elephants and Man

It takes an elephant approximately thirteen years to mature and reach its full adult size. Until then, they are not good pack or draft animals, and they require enormous quantities of food. Because of this, elephants are rarely bred and raised in captivity (except in zoos and in India where the sacred white elephant is found). Instead, adolescent elephants are captured, tamed and trained. An intelligent animal, the elephant readily learns some twenty commands. It also teaches its keeper to recognize a few of its own commands—the demand for food, for example. And it is capable of forming strong ties to its trainer and showing great affection. Wild elephants living in preserves grow accustomed to human presence and will often become friendly with gamekeepers and even visitors.

The Falcon

Like the flight of gryfalcons from
their native haunts . . .

The gryfalcon is a large Arctic raptor, and while the poet Heredia may have found its power and predaceous habits an apt metaphor for those of the Spanish conquistadores, there the behavioral similarity ends, since the gyrfalcon rarely migrates from its Arctic haunts. Only occasionally, in winter, is it seen as far south as Maine and central Canada.

Perhaps its rarity made it the favored falcon of English royalty who used these birds to hunt small prey. In fact, the sport was so popular that by the species of bird a man carried on his wrist one could practically determine his position in society. The peregrine falcon was carried by an earl, the goshawk by a yeoman, the merlin by a priest and the smallest bird, the kestrel, by a servant. Falconry has largely disappeared as a sport, and where it is still practiced it is the subject of increasing controversy, not only because it is a blood sport but because several of the falcon species used are now very scarce in the wild.

It is not surprising that falcons should aid man in hunting, since the falcon has extremely keen vision. It is swift and skillful in capturing flying prey—very few escape the bird's plunging dive. Peregrine falcons have been clocked at over two hundred miles per hour. Falcons can also be taught to hunt specific birds and are trained to recognize which prey species to attack on hearing a vocal command from the falconer. The origins of the sport of falconry probably trace to the deserts of the Middle East and North Africa. It spread to Europe during the Crusades and reached a peak of popularity in medieval England. Today the last stronghold of falconry is still the desert of the Mideast where sheikhs and their families raise and train their birds. It is not unusual to see a sheikh driving his Cadillac into the desert for a day of hunting with his birds.

Perched on the falconer's arm.

The Language of Falcons

Falcons give hardly more than a single cry during most of the year. This cry varies according to the species: "killy-killy-killy" for the kestrel or sparrow hawk, "klee-klee-klee" for the merlin or pigeon hawk and similar cries for the others.

However, during the mating season the birds become very talkative and use a series of strident cries which sound like "grae-grae-grae" or "guu-guu-guu." Near the nest of a kestrel, the "killy" cry is frequent and is heard in towns where these birds perch on towers and steeples. As recently as forty years ago, kestrels and merlins were fairly common in New York City where they nested on ledges of buildings along Fifth Avenue and Central Park West, high above Central Park. From these aeries, they could easily watch for and prey on the pigeons, sparrows, rodents and other small animals which inhabit the park. While they have practically disappeared from this exclusive neighborhood—partly because of pollution and population density and partly because they were simply dislodged by human neighbors—they still fly over and stop in New York regularly during the spring and fall migrations, and their slow circling is often seen by office workers on upper stories. In Brooklyn, in 1978, a pair of kestrels nested and raised their young on a window ledge of a brownstone, much to the delight of their human hosts.

During courtship, like all birds of prey, the male falcon performs a flying display which includes an extraordinary spectacle of aerial acrobatics. First, he gains great altitude, then dives in a dizzying plunge. He brakes forcefully just above the female and, wings spread wide, he hovers there, beating his wings to stay in place and uttering the strident cries described above.

Fireflies and Glowworms

During summer evenings, lawns, gardens and even some city parks are filled with the little flickering lights of fireflies or the greenish glow of the glowworm moving in the grass. These insects have long fascinated humans. Luminescent species produce their light by releasing minute amounts of an enzyme called "luciferase" which reacts with a biochemical pigment called "luciferin" to create the cold light or glow. These same two substances are found in other luminous creatures, such as certain deep ocean fish and worms.

The Light Is the Message

The light of the female glowworm attracts the male, who remains dark and practically invisible. Unlike the glowworm, fireflies of both sexes signal and respond to each other with their light. However, for the most part it is the brighter light of the male which we see, while the female perches in the shelter of a tree.

There are several species of insects capable of luminescence, and each one has its own pattern and rhythm of flickers. Color and frequency are important. Occasionally a multicolor combination is used—a reddish light will be followed by a greenish one, for example. These codes are precisely and rigidly defined and the slightest change will cause the light to be ignored. For example, one species will respond to flashes at a sixteenth of a second but will ignore those at a twentieth of a second. A male glowworm, attracted by a brilliant but artificial light will approach it with interest until he is close enough to realize his error in size and then will simply turn around and retreat. However, luminous glowworm lures are sometimes successful if the light source is of the proper size and intensity. Paradoxically, a source which contains more yellow light than that normally emitted by the female glowworm will attract many more males, but no one yet knows quite why. Nor is there any explanation for the synchronized flickers of entire colonies of certain fireflies in the tropics where these luminous insects are abundant.

Man is not the only one to attempt to imitate the signals of luminous insects, nor is he as successful as the female photuris, a tropical insect. This natural predator uses her own luminescence to imitate precisely the signals of the females of several other species and attract the males, which she quickly kills and eats. She does not use her light to attract her own mate.

Fireflies flashing synchronously.

The Fox

The fox has managed to survive centuries of fox hunts, commercial fur trappers and other more lethal programs of extermination. Gas, poisoned bait and other traps have been used, and bounties have been offered in some places for fox carcasses. In France alone, more than 2 million foxes have been exterminated in the past ten years. The fox has also been threatened by natural pressures. For example, in New England, they almost completely disappeared when a particularly virulent disease attacked pups and adults alike in the 1960s. They are making a comeback but are shy of humans, and many of the former haunts have since given way to human development.

Communication

Only recently have foxes been the subject of serious study, and we still know little of their habits and communication. We do know that they are not as solitary as once believed. Not only do they stay together during mating and raising their young, but at other times they use their scent to signal to other foxes in the area. One way in which they do this is by scent-marking to identify food. A fox is attracted to all sorts of waste products—bones, fallen fruit, garbage, carrion and so on. It first satisfies its own hunger and then, when it has either finished or determined that the food is

inedible, it deposits a few drops of urine on the leftovers. Other foxes coming across the same food will almost certainly ignore it once they smell the urine left by the first fox. The fox has a very strong odor that even humans can smell and it uses this powerful medium to communicate.

Conversation

The vocal language of the fox is not well understood. We do know for certain that males and females make slightly different basic yelping sounds and that the female has a "grrou" cry of alarm which she uses to alert her pups. She also makes a clucking sound while caressing or playing with the pups. The male fox makes this same sound when bringing his kill to the den. The males also make grunting sounds which have yet to be interpreted.

The pups' cries resemble those of angry kittens or canine puppies. Some cries of adults resemble those of owls, so much so that one animal has been known to respond to the other on occasion. This may be unintentional mimicry, as may be the fox's bisyllabic cry which resembles the cry of the peacock. Dogs in the area will respond to these peacock-like calls.

The Gannet

This fish-eating seabird is both handsome in appearance and a spectacular performer. During migration it flies seemingly without effort above the sea, from time to time plunging straight down into the water from heights of fifty feet or more to spear a fish with its strong beak. It is protected from injury when it hits the water by a cushion of interconnected air sacs under its breast skin.

Gannets nest in enormous colonies in far northern latitudes, notably on islands off Newfoundland, Nova Scotia and in the Gulf of St. Lawrence. Over 15,000 pairs form the colony on Bonaventure Island in Quebec. Other colonies are found on the northwest coast of Europe, particularly in the Hebrides off Scotland. Within these extraordinary colonies, each breeding pair occupies and defends a small nesting space—only about 2½ feet square. Preferred locations are on the sides and ledges of precipitous cliffs, but they will also build their seaweed nests along the top of the bluffs. If a bird lands on a neighbor's nesting space, turmoil erupts and the intruder is soundly slashed with the gannet's strong, pointed bill. A prolonged fight can mean serious damage, perhaps the loss of an eye, to one or the other adversary, but most fights begin and end quickly with nothing more than a sharp peck inflicted. Most of these fights are territorial and occur between two males.

The natural uproar made by the colony even in its

peaceful state is almost deafening, yet each bird can recognize the guttural croaking or grunting sound made by its mate or hatchling. These sounds, made only during the breeding season, are also the only ones ever made by the gannet, which is without a proper song. But, like other members of the booby family, the gannet has an elaborate system of displays and gestures.

Once the male has staked his claim to a nesting site, he spreads his wings and lowers his head and neck in a threat display which warns away other males in search of nest space. This same position also attracts the female, who recognizes it to mean that the male has found a nest. She approaches cautiously, tilting her head to the side and offering her neck which the male takes gently in his bill to signify his acceptance of her as a mate. Both birds then raise their heads and cross bills, not unlike human fencers; they bow with wings and tails raised. This fencing and bowing behav-

Body language of the gannet. 1. Threat display. 2. Female approaches (left) head tilted. 3. Crossing and clapping bills. 4. Preparing to depart the nest.

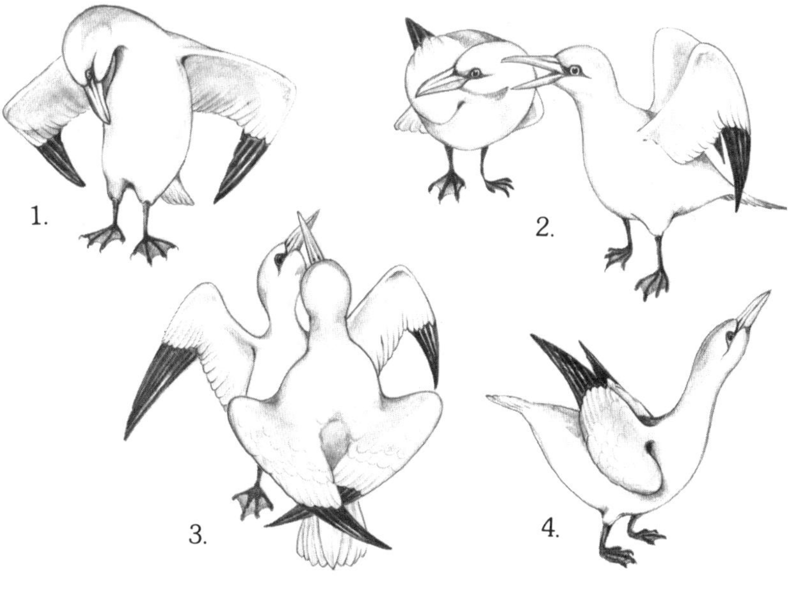

ior is repeated each time one or the other returns to the nest. The greeting is followed by mutual preening of head and neck.

Once paired, the female lays only one egg which she and her mate take turns incubating. Later they share the duties of tending the nest and raising the hatchling. The two take turns flying off to feed. The returning bird, after performing the greeting ritual, will also collect small sticks and bits of seaweed and present these to the bird about to be relieved of nest duty. The departing bird will stand briefly with head stretched straight upward and wings raised over its back to signify that it is leaving. From this position, it launches itself into the air with a vertical leap.

Goats and Sheep

These two species were domesticated in Mesopotamia as long ago as 5000 B.C. and are mentioned in several places in the book of Genesis. In one passage, Abraham's fortune is reckoned: ". . . some sheep, oxen, donkeys, male and female servants, she-asses and camels." The order of the list is interesting to contemplate, inasmuch as sheep are given precedence over servants. Perhaps this should not surprise us, since over the centuries servants have become increasingly rare while sheep and goats continue to provide us with wool and excellent cheeses.

Although sheep differ from goats in many obvious respects, the similarities in some of their behavioral traits—particularly their systems of communication—are striking. Both species use scent signals. Not only do odors play a part in courtship displays between sexually mature adults (male goats particularly give off a very strong odor during the breeding season), but they help establish and maintain relations between mothers and their young.

Sheep also depend heavily on vocal communication. Bleating is usually a contact cry made by a lamb or adult sheep separated from the flock. Among lambs and kid goats it is also, when given more sharply, a distress cry to alert the mothers that the young animals are upset, lost or otherwise distressed.

During the breeding season, the male of both species makes a characteristic low grunting cry with which he summons the female. Another cry, "m-m-m," is made by the adults of both species to establish contact with their offspring.

Rams fighting.
Male goat making
an extreme threat display.

The animals' responses to danger or threats differ somewhat: In giving alarm signals, sheep snort and stamp the ground; goats do the same but much more sharply and frequently. The differences between the aggressive postures of the two species are more pronounced. During a confrontation, two rams assume a frontal threatening posture and fight forehead to forehead. Two goats, on the other hand, stand on their hind legs, each one thrusting himself forward and downward onto his opponent, trying to butt him with the side of the head. The aim of this jowl-to-jowl move is to lock horns.

When goats and sheep graze together in a single flock, fights between the two species are rare, and when they do occur they are over almost before they begin.

One thing goats and sheep do not share is temperament. Goats are relatively intelligent animals capable of escaping from enclosures and exhibiting a natural curiosity. It is not unusual for suburban goats, kept partly as pets, to find their way into a kitchen, given the slightest opportunity. Sheep would never behave this way unless led. They are more passive and submissive and would not spontaneously attempt to find a way out of their pasture.

The Goose

The domestic goose, *Anser anser,* is descended from the wild greylag goose. In general, geese and other waterfowl are easily tamed and were probably among the first wild creatures to be domesticated. We know that the Latin name *Anser* is derived from the more ancient Sanskrit name *Hansa,* and we can deduce from this that the goose has been raised almost since prehistoric times. Their images are found in ancient Egyptian hieroglyphs and tomb paintings, and in the art of other ancient civilizations throughout the world.

Conversation

Like ducks to whom they are closely related, wild and domestic geese share many of the same behavioral characteristics, including a common language with slight variations in accent. For example, the greylag's cry is the classic honking, while the laughing goose makes a sharper, more raucous sound.

Geese have a single basic cry, the honk, which actually sounds more like "gahng." With this one type of sound the goose can express several different things. Geese are constantly honking to each other, whether in flight, on the ground or on the water, relaying information about the choice of resting place, the quality of the weather, the best grazing sites or the presence of hunters or other dangers. Even at rest, grazing on a prairie, geese continue to honk softly in series of six or seven repeated "gahngs." As the time approaches for the flock to take flight and depart, their cries grow louder and the repetitions fewer. They begin to walk or swim more rapidly, signaling their intention to fly off. Flight usually begins when the cry has been reduced to two honks—"gahng-gahng."

A single loud "gahng" is usually a warning cry easily recognized by the rest of the flock. Alerted to the presence of danger, the male and unpaired adult birds take flight immediately, while the young goslings rush to their mothers who find some cover for them to hide in until the danger passes.

Body Language

In every grazing or resting flock of geese there are one or two birds which stand sentinel over the rest. When approached by humans or other intruders, they will charge, honking and hissing loudly, neck extended, wings spread to increase their apparent size. If the intruder does not withdraw, the threat display can become an actual attack; geese are capable of inflicting serious damage to dogs, cows, sheep, humans or anyone who trespasses in their territory. Their beaks, necks and powerful wings make formidable weapons.

The male goose adopts this same threatening posture when courting his mate, but in this case it is usually only a display and rarely escalates to anything more serious. The male will circle around before reaching the intruder who is the object of his actions and will return to his mate. She will greet him by honking and bowing, raising and lowering her head and neck. He will bow and honk in response.

This ceremony serves to reaffirm the bond between male and female. Many geese, including greylags, mate for life and will not choose another mate if their partners die.

Domestic geese are strongly territorial and every bit as

Courtship and threat display of the greylag goose.

alert in guarding their homes as the reliable watchdog. This quality was discovered as long ago as 390 B.C., when flocks of geese were confiscated by the Romans in Gaul and driven like cattle overland to Rome. Not long afterward the geese raised a cry of alarm as a group of Gallic invaders were about to attack. Some have suggested that they may have been greeting the breeders who raised them, but it is much more likely that they were responding to the threat of "invasion" of their new territory.

Perhaps the most fascinating example of man-to-goose communication is found in Konrad Lorenz's study of the imprinting behavior of the greylag gosling. Like ducklings, goslings will attach themselves to the most suitable mother figure available if their own is not immediately nearby at hatching time. Lorenz allowed a brood of goslings to imprint on him. He made every effort to imitate the behavior of the mother goose and would waddle along with his charges waddling behind him. They responded to him vocally and visually just as they would have to their real mother. He taught them to swim and, with great difficulty, he actually taught them to fly. Much of his early work was documented on film and in highly readable books which are as amusing as they are informative.

The Great Crested Grebe

This bird is a native of Europe and only rarely seen in the United States. However, its habits are similar to American grebes and all members of this family carry the distinctive ruff or crest on their heads during the breeding season.

Communication

The breeding season begins in March with the highly ritualized courtship ceremony. Face to face, the male and female swim toward each other. When they meet they stretch out their necks, shake their heads vigorously and then swim peacefully side by side for a short distance. This mutual display is repeated five or six times, accompanied by a peculiar barking cry composed of two notes. The two birds then dive together repeatedly. They surface, holding weeds in their beaks which they offer to each other. Such offerings are fairly common among birds. Gannets, bowerbirds and many others have been observed offering plants, insect larvae, fish, seaweed, shells and other such gifts. This is clearly a symbolic gesture, for the gift, if edible, is rarely eaten.

Courtship display among grebes. Top: face to face. Bottom: mutual offerings.

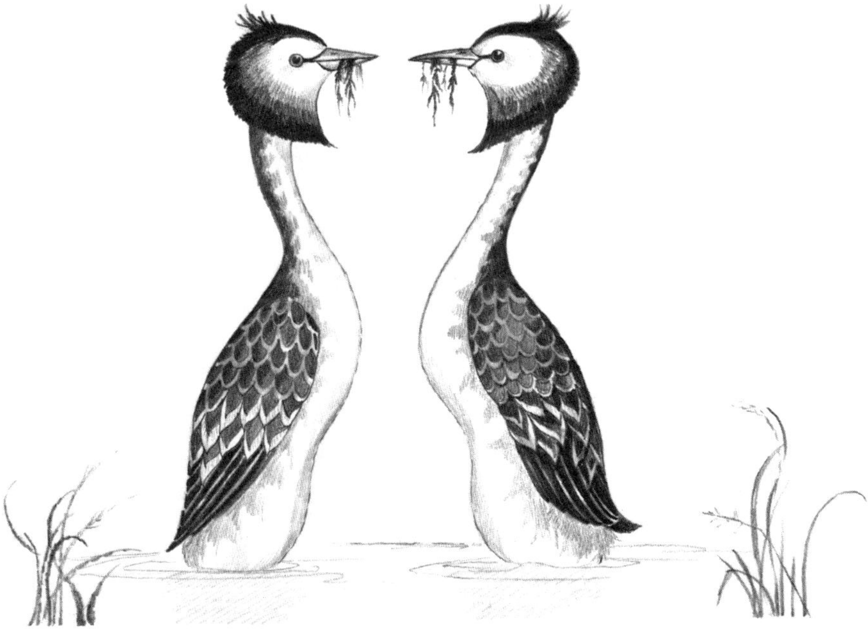

The Guinea Pig

The Incas of Peru first domesticated their native guinea pigs as a food source. Today guinea pigs, still abundant in their natural habitat, are also abundant in laboratories and in homes where they are kept as pets. Years of selective breeding have resulted in several varieties, such as the long-haired Peruvian and the Abyssinian rough-coated guinea pig. All are tailless and most average about 1½ pounds in weight.

Conversation

The guinea pig strikes us as fairly quiet, but some eighteen different cries and grunts have been identified. They take on very expressive postures and have a distinctive odor by which they recognize familiar guinea pigs and isolate strangers. The basic and most frequently heard cry is a whistling sound. When angry, the male raises his head high, makes muffled noises, then whistles. He also grinds his teeth to produce a screeching sound. If provoked he stretches his head forward to confront his adversary muzzle to muzzle before actually biting. When bitten by another guinea pig or otherwise hurt, he makes a sharp squeal not unlike that of a pig.

During courtship, the male investigates the genito-anal region of the female. At the same time, he purrs somewhat like a cat. If the female is unresponsive, she utters a sharp, short squeak and may then turn and bite the unfortunate male.

Baby guinea pigs make a series of very sharp, high-pitched little cries grouped in twos or threes when disturbed. The same cry is used by the mother as a response and by strayed adults of both sexes as a contact cry.

Top: Attacking posture. Middle: Muzzle-to-muzzle confrontation. Bottom: Neutral position.

Humans and Guinea Pigs

Because guinea pigs are so frequently kept as pets, they offer a good opportunity for amateur observers who are willing to watch patiently and interpret the signals, many of which are quite clear. An unmistakable whimper cry, repeated incessantly and rising in tone is used to demand food. As with the cat, a purring sound usually means contentment and may be heard when the guinea pig is fed some particularly favored food such as fresh parsley or lettuce. They will occasionally purr when their bedding is changed, when they are given the opportunity to run in the grass or when they are stroked.

After centuries of domestication the guinea pig has proved that it adapts easily to a human environment. However, females are sometimes easier to tame and handle than males, and a solitary guinea pig will be more interested in humans than will one with companions.

In one case, a boy established a firm rapport with his guinea pig. He would click his tongue and his pet would actually wag its ears and purr briefly in response, eventually turning its head and coming over to be stroked. Another pet guinea pig had a weakness for the fringe on an old Persian carpet. When allowed to run loose, it would nibble along the edge and if caught would adopt a submissive posture. Not only are pet guinea pigs capable of learning which actions are allowed and which—such as eating the carpet—are forbidden, they will respond to a given name and will readily come to recognize the different people who make up the household.

The Guppy

This freshwater species of fish inhabits the waters of the northeast corner of South America and the Antilles. It is probably one of the best-known aquarium fish and even in the care of the most inexperienced amateur will usually thrive and breed readily. In fact, it is so prolific that it has earned an appropriate alternate name: millions fish.

The male's bright rainbow coloring, especially on the caudal or tail fin, is in striking contrast to the more subdued grayish coloring of the female who is slightly larger. Guppies are tranquil and sociable, rarely fighting with other species. Because they reproduce with such ease in captivity, they make fascinating tank fish.

Courtship display, male flattening his fins.

The male initiates courtship by following close behind the female, simultaneously closing his fins. The tail fin is drawn so tightly closed that it seems pointed, making the male appear smaller and less threatening to the female. When he has engaged the female's attention, he places himself in a perpendicular position across her path blocking her way, keeping his fins flattened.

The brilliant spots on the male's fins are also a threatening signal. When two males confront each other, the fins are spread as wide as possible in a threat display designed to exhibit the color spots completely and drive off the adversary. In the event of actual combat, the two challengers swim parallel to each other, hitting out with their tails.

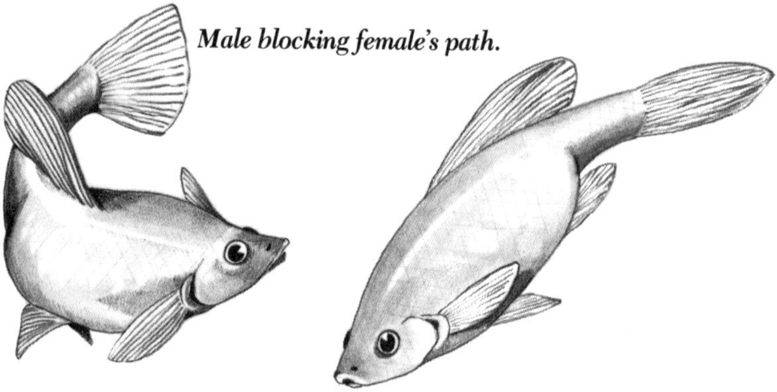

Male blocking female's path.

Fighting position adopted by two male guppies.

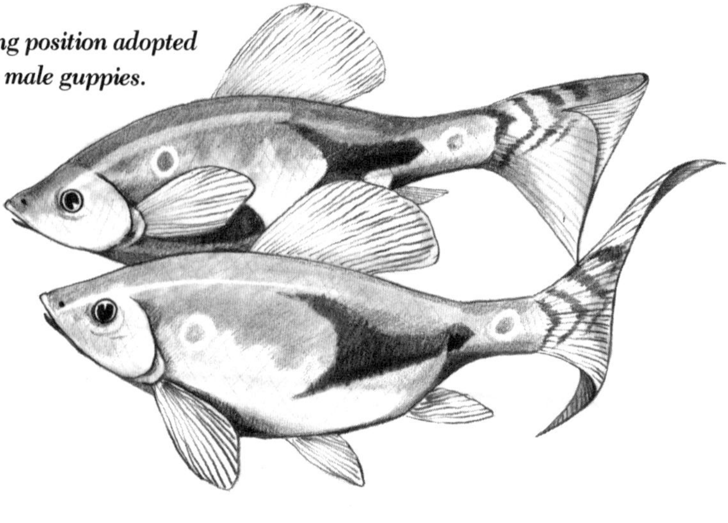

The Hedgehog

This small nocturnal mammal earned its common name from its habit of frequenting hedgerows in developed areas. It is a native of Europe, Africa and Asia, and in the United States can be found only in zoos and occasionally in homes where it may be kept as a pet.

One of the more interesting habits of the nearsighted hedgehog is its defensive posture. Its body is covered with coarse hairs, some of which evolution has modified into sharp spines. When disturbed or frightened, the hedgehog rolls into a tight ball and activates special subcutaneous muscles which erect the spines, making him a prickly package for any would-be predator.

While hedgehogs are primarily insectivorous, they also feed on birds, eggs and small mammals when they can catch them, and they will browse contentedly through the garbage can for food scraps. Although they have adapted fairly well to the presence of humans in their habitat, they often fall victim to automobile traffic. At night—especially during the summer—they are attracted onto roads by the heat retained by the asphalt or concrete and will crouch down to warm themselves and aid their digestive processes. This habit, combined with their nearsightedness and passive defense, accounts for their poor showing in avoiding oncoming cars. In France, where hedgehogs are plentiful, it is estimated that from forty thousand to fifty thousand are killed this way each year.

Typical foraging posture of the hedgehog, guided primarily by scent.

Although hedgehogs have a complex variety of cries, wheezes and grunts, they communicate with each other primarily through scent, smearing their bodies with saliva to enhance their own odor. First it chews small pieces of food, producing a frothy, highly odorous liquid which it spreads over belly, flanks and paws and then leaves in small traces along its route as it travels, indicating to other hedgehogs that it has passed. This saliva-enhanced scent is also used as a contact signal between mother and young who use it to follow her more easily.

Hedgehogs and Humans

Apart from their often disastrous sallies onto highways, hedgehogs are fairly shy animals. Occasionally they have been kept as pets and then have proved to be responsive to human contact, within limits. One family rescued a badly dehydrated hedgehog during a drought. It was too weak even to defend itself by curling up. After ridding the animal of external parasites, which were contributing to its poor condition, it was given drinking water and the run of a bathroom. Within a few days it had recovered sufficiently to eat and was happily consuming scraps of sausage, cheese, eggs, fresh fruit—and had developed a particular liking for cottage cheese. It soon felt at ease and gradually stopped curling up whenever anyone approached. Eventually the hedgehog even came to greet its human hosts with cries of welcome.

Notwithstanding this rewarding encounter between the family and the hedgehog, which was finally returned to the wild, it is usually safer to observe this wild animal in its natural habitat, or at most when coaxed into the open by leaving scraps of food near the garbage can. The hedgehog usually approaches noisily and can be heard at some distance, rustling leaves and breaking branches in the underbrush.

The Horse

Equus caballus, the horse. Perhaps more books have been written about the horse than any other domesticated animal. The horse has shared man's history, aided him in his work, accompanied him into battle and carried him into the wilderness. It has made men's fortunes at the race track and inspired him in his creative life, figuring prominently in mythology, literature and the visual arts. The horse has been treated with reverence and respect, but like other domestic animals, it has sometimes suffered under man's guardianship, and until recently the horse's behavior, except insofar as it affected human interests, had been largely ignored. While we know a great deal about training, breeding and keeping horses, we know little about why they behave as they do with their own kind, and how they communicate.

The Herd

A herd of horses, whether in the farm pasture or on the open plain, is organized into a social hierarchy. The dominant animal or leader is usually a powerful stallion, unless the herd consists only of mares. In that case, an older mare will assume the dominant role. Beneath the leader is a descending order of individuals—the older, heavier horses, also usually stallions, in the highest rank and the younger, weaker horses in subordinant positions. The hierarchy is established and maintained by a series of challenges and duels, with the younger, subordinate horses usually initiating the combat. Higher ranked horses must fight harder and more frequently to maintain their rank against these challengers. The dominant horse in the herd is always alert and will discourage challenges with an immediate show of strength against the advances of other herd members. Animals of the same rank will also fight each other.

During combat, the animals show various types of aggressive behavior, including biting, pushing and kicking, which will be discussed below. A tamed horse kept by itself in a stall or paddock, not having to fight for its rank in the herd, will become aggressive only if it is frightened. This behavioral response cannot be overcome with harsh punishment, which will only make the horse even more aggressive. Instead, a firm but gentle hand and soothing voice will more readily dispel the horse's fear and calm it down.

Conversation

The familiar neighing or whinnying of the horse is often either a distress signal or a contact cry made by a horse when it has been separated from the herd. When softly made, it is also used as a greeting. Snorting is a signal of danger, used to alert the herd. Snorts can be heard for some distance, even for miles in the wild, and are effective warning signals.

Body Language

The horse uses small muscular movements of the legs, ears, lips and skin around the nose to relay messages about its state. The signals are numerous and varied and we have yet to definitively interpret many of them. We do know that a large number of discrete signals exist and that the horse is extremely sensitive to visual and tactile stimulation.

In 1911, a horse named Clever Hans astonished the scientific community when he demonstrated an apparent ability to learn complex arithmetical skills and to transmit the results of his computations by tapping one foot. Although many scientists, including the experimenter, were convinced that the accomplishment was real, more careful study revealed that the horse was actually watching the person presenting the problem who would invariably make subtle unconscious movements while counting the horse's taps in order to determine whether he had reached the correct answer. The human observer would hesitate, stop or make some expectant sign—again unconsciously—when the correct number was reached, and Hans simply learned to stop tapping when he saw this signal. The experimenter had drawn the wrong conclusions, and the horse was performing in a much simpler way than was thought. Even today similar misinterpretations of behavior are called "Clever Hans errors."

But one thing was clearly revealed: The horse is capable of recognizing and interpreting very subtle visual signals, capable too of recognizing and responding to equally subtle tactile stimuli. The art of dressage demonstrates this sensitivity. The horse is trained to perform complicated movements and maneuvers when touched very gently by the rider in specific places on the horse's body. Often the signals are so subtle that the untrained observer is altogether unaware of them.

But dressage is a language designed by man, not horse. The language the horse itself uses to communicate with humans or other horses is often just as subtle. Several, though by no means all, of the gestures have been interpreted. For example, domesticated horses lay back their ears as a warning not to approach. If the signal is ignored, the horse may begin to wave its tail like a whip, meaning that it is preparing to kick. A raised head and protruding upper lip may mean that the horse is ill, although there are other possible causes for this expression. Horses moving together in a herd will fall in behind the dominant animal and travel in single file, each with its nose to the tail of the animal preceding it, ready to respond to whatever signals are passed along. The leader alone is responsible for keeping watch, his head held high, his eyes fixed on the herd's destination.

POSTURES OF ALERT AND ALARM. When a horse discovers an intruder, it assumes a posture of alert. It raises its head, the ears swivel quickly from front to back listening intently for sounds from the intruder or the rest of the herd. It stands perfectly balanced, weight distributed evenly. Its tail is slightly raised. The rest of the herd will recognize from this posture the presence of possible danger, but the alerted horse will neigh all the same to warn any who haven't yet noticed. If lower ranking members of the herd are the first to notice the intruder, they will neigh to attract the attention of the dominant mare or stallion. If the intruder is a human whose presence does not disturb the dominant animal, he or she remains calm and continues to graze and the rest of the herd quiets down, twitching their nostrils—an equine sigh of relief—to indicate that they are reassured.

If the dominant animal *is* disturbed after its inspection of the intruder, it snorts loudly. The other horses then adopt a posture of alarm, which is a more intense posture of alert. Head, neck and tail are held higher. The dominant horse takes several quick steps toward the intruder for a closer evaluation and attempts to stare him down. Then he snorts again and the other horses move to stand behind him, all facing the intruder. The herd's response to the threat is still uncertain at this stage and several of the horses may raise their tails and defecate, signaling their uncertainty.

FLIGHT BEHAVIOR. If the dominant animal determines that the intruder is dangerous, it will signal the herd to retreat by turning its back on the intruder. The herd, shifting around, will follow closely behind him. Usually they move no more than fifty to one hundred yards before stopping to observe the intruder again. The intensity of their alarm depends upon such factors as the sensitivity of the group, the proximity of the intruder and the degree of surprise.

AGGRESSIVE BEHAVIOR. Horses prefer to retreat from an outside threat rather than fight, but if retreat is impossible they will defend themselves, sometimes even taking the initiative, whether their adversary is another horse, another species or man.

Fights between horses regularly occur when one challenges the hierarchical position of another. There are two basic fighting styles—one based on biting, the other on kicking—but both include a prelude of increasingly menacing postures. Before biting, the horse will flatten its ears against the upper part of the neck while its head is turned toward the adversary. Its mouth is open, with the lips pulled vertically to expose the incisors and sometimes the canine teeth as the horse finally thrusts forward to bite.

Before kicking, the horse turns its back on the adversary and begins whipping its tail. It raises and tucks one hind leg beneath its hindquarters, then kicks the hind leg in the direction of its victim. These blows can be extremely accurate. It is also capable of kicking with both hind legs. In a single motion, it jerks both legs under its hindquarters, throwing its weight onto its front legs and straightening its hind legs in a devastating kick.

Sometimes, kicking and biting are symbolic threat displays to discourage challengers or show strength. This is often the case with young colts or lone stallions.

Whether the fight consists of biting or kicking, the horse shifts its ears, head and tail continuously. At the provocation stage, ears are held forward, head high, neck arched, mouth closed and tail raised. When the fighting escalates the action speeds up. Such fights between two horses include a great deal of violent, noisy jostling and butting. To avoid a fight, a submissive horse holds its ears upright and slightly for-

1. Posture of alert. 2. The horse moving to withdraw. 3. Threat display by one horse (right) to another—ears laid back, head rocking, about to bite. 4. Mare (left) refuses the stallion and is about to kick to drive him away.

ward, mouth closed, head and neck turned aside to avoid the rival's bites.

During the breeding season fights frequently break out between stallions. While a fight is in progress, the mares ignore it, continuing to graze but remaining in a close group awaiting the end of the match. Colts and often stallions sniff each other's genital areas before fighting. This is a social ceremony, the male of superior rank being attended to by the inferior male.

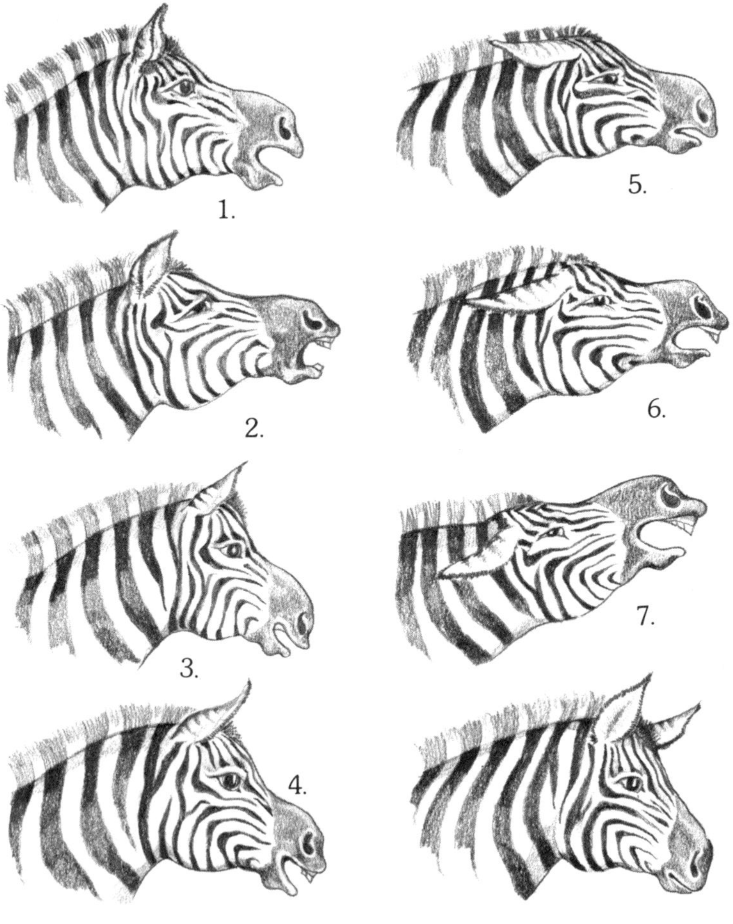

The facial expressions of the zebra closely resemble those of the domestic horse. 1. Hunger. 2. Yawn. 3 and 4. Friendly greetings. 5 — 7. Threat displays of increasing intensity. 8. Alert and attentive.

The (Blue) Jay

This handsome large blue bird is found throughout the United States east of the Rocky Mountains. It is a member of the crow family, *Corvidae,* thought by many to be the "most evolved" of the bird families with different species found around the world. Many are omnivorous; they can be either gregarious or solitary and extremely talkative; they have also adapted well to the presence of man and, while they prefer oak forests, they are found in city parks and suburbs.

Conversation

The jay has an extensive repertory of individualized cries and calls. It is also capable of uncannily accurate imitation of the songs and cries of other birds and of such miscellaneous sounds as the barking of a dog or a human whistle. The mimicry is often purer and sweeter sounding than the original.

The jay's summoning cry (Figure 1) is a single plaintive note, "eeeh," resembling the meowing of a cat. The same cry is used as a contact cry between adults and sometimes as a hunger cry by the young, even though they have a special *feeding cry* which resembles the summoning cry but comes from deep in the throat and is more insistent. It is heard when one jay returns to feed its chicks or mate.

Feeding is often part of the courtship ritual. Usually the male will begin by offering food to the female in response to her feeding cry. This behavior may be preparatory to the cooperation required later when the eggs are laid. Or the roles may be reversed, with the male summoning the female to the future nesting site and demanding food (see Figure 2). If the male is weak and in feeble health, the female will provide him with food.

During spring mating, there is a special loud summoning cry which resembles the "kraah-aah" of the crow. It has variations such as "kyankh" or "kyarnk." This is a sexual call used by females to attract the male.

Warbling and chirping are a series of soft stuttering notes, including variations of the summoning cry but louder and more prolonged. There are many variations, depending on the individual bird. Every note is accompanied by a movement of the body and tail; the bill is wide open when the bird sounds the first long note. Warbling is used exclusively by paired birds.

The cries of distress and alarm are both calls which are more piercing, raucous shrieks usually repeated twice in rapid succession. The bird stands upright, feathers smooth and held tightly against the body. Then, as the bird sounds the alarm, the head and body bob up and down and from side to side. Occasionally the crest is raised (see Figure 3).

Other birds recognize the jay's distress cry to mean that an intruder or predator is near. Very often the jay will sound the alarm as it retreats from a predator or when it is trapped or captive. (Caged male blue jays, especially, have been observed calling incessantly.)

The blue jay tailors its various cries of alarm to the nature of the threat, and here again the other birds of the forest recognize the jay's intention. It uses a "meow" type of call to signify hawks and buzzards, "kik-kik-kik-kik" for a kestrel, and it can almost exactly mimic the whistled scream of the red-shouldered hawk. For a squirrel the jay makes a series of short cries, and for a human one sharp note. Except for this sentinel service provided to other birds, the presence of jays is unwelcome to most other species. Jays prey on eggs and hatchlings of smaller birds and are notorious nest snatchers.

The cry of aggression is a curious short, grating cry, resembling the raucous clicking of the thrush, which the jay makes when attacking, defending itself or quarreling with another jay. Still another cry is the "tscherr" cry which still puzzles ornithologists. It is monosyllabic, very loud and sudden, sounding somewhat like water rushing out of a faucet.

Giving cry of distress or alarm.

The passive ant bath.

The body language of the common jay.

Male summoning the female.

Threat display, crest feathers raised.

The female's submissive posture.

Hunger demand of an adult jay resembles that of a hatchling.

Body Language

The male jay's threat display is designed to make it appear larger (see Figure 4). It ruffles its feathers and raises its crest to intimidate a rival and attract a female. The female may also assume this pose to drive away rivals or to signify her acceptance of a male. Often however, the threat display is the prelude to actual combat.

Battles occuring between two pairs of jays are probably related to territorial disputes. The pair in residence will direct threat displays toward the two trespassers. If the intruders do not retreat they will be attacked and ejected. Occasionally a bird will be afraid to attack its adversary and will actually transfer its anger and aggression to an inanimate object such as a stone or twig.

Courtship

To indicate her interest in a male, the female jay rocks her body and waves her tail up and down. Then she raises her head very high, moving it slowly from side to side, exposing her breast plumage. Once paired, female jays continue to greet their mates in this manner. They converse in sweet warbling until the female adopts a submissive posture, her trembling wings spread wide and drawn back, tail waving and rocking (see Figure 5).

The Ant Bath

The bathing behavior of jays is unusual. Not only will they splash in a puddle like any other bird, but they use ants as a "dry cleaning" agent. Jays have been observed near anthills, wings spread on the ground (see Figure 6), allowing ants to crawl over their feathers and skin. Other jays catch an ant in their bill and rub its body over their feathers. This seemingly peculiar behavior is for good reason. Ants are rich in formic acid which kills the parasites living on the jay and helps to polish the large quills. So useful are ants that jays, which eat all manner of insects—beetles, fleas, larvae and others—will not eat them, although they do consume the winged ants which do not contain formic acid.

Jays and Humans

Blue jays are best observed in their natural habitat, but like other members of the *Corvidae* family, they are easily tamed. They can be befriended through regular feeding but should not be captured. Once a jay comes to recognize a familiar human, it may adopt a posture of submission while being fed and will respond to the summoning cry and cry of alarm if they are well imitated. Female jays may even present the spread wing display she makes before her mate and allow herself to be stroked, but this is unusual even in caged jays. A tamed jay will demand to be fed just as a young hatchling would, beating its wings and twitching its tail.

The Kangaroo

The kangaroo is a marsupial, a rather primitive mammal which bears extremely underdeveloped young. Lacking a placenta, the female raises them in an external pouch that contains the mammary glands. Most marsupials, kangaroos included, are found in Australia and South America, although the United States has one native species, the opossum. However, aside from their method of raising their young, opossums and kangaroos have little in common. There are several species of kangaroo, some reaching a height of four or five feet when standing upright. All are characterized by short forelimbs and large, powerful hind legs well adapted to the leaps for which the kangaroo is famous. Their long, tapered tails are quite strong and are used for balance.

Conversation

Until it can fend for itself, the young kangaroo spends most of its time in its mother's pouch, attracting her attention by leaning its head out and making a "mee-mee" cry. The mother kangaroo responds with a clucking sound which reassures and calms the infant. By the time the youngster is ready for weaning, its cry resembles that of the female, though somewhat sharper in tone. With this cry, mothers and young call to each other loudly and constantly as they graze so that they do not lose contact with one another. The male kangaroo's voice deepens as he matures. He seeks a mate by following close behind an available female and making a low, soft cry.

Kangaroos travel together in packs, foraging and grazing. They are organized into a social hierarchy and several of the larger, dominant males will stand guard, alerting their companions to the presence of danger with an odd, noisy explosive cough. The startled kangaroos jump much higher than usual, landing heavily and repeatedly slapping their rear paws against the ground. This further alert usually signals the withdrawal of the entire pack to some distance.

When threatened, the older, stronger males adopt an upright menacing posture that resembles a boxer's stance. Kangaroos fight each other by kicking with their hind legs as well as striking blows with their weaker forelimbs.

Fighting posture of adult kangaroos.

The Lobster

While not as visible as mammals, birds and other vertebrates, arthropods (invertebrates characterized by jointed legs) comprise almost 90 percent of all living creatures. This staggering figure may seem more reasonable when one realizes that insects, crustaceans, arachnids (spiders) and myriapods are all arthropods. And of all the arthropods, the sea-dwelling lobster is the largest. Only a few species inhabit American waters, and of these the East Coast Maine lobster with its two large front claws is most familiar. It is found from the coast of Labrador to North Carolina, but is most prevalent in a fifty-mile-wide strip off the coast of Nova Scotia and Maine. Most of those caught are about five years old and weigh three to five pounds, but not long ago a lobsterman pulled in a specimen weighing forty-seven pounds and estimated to be almost fifty years old.

Like other crustaceans, the lobster has a hard exterior shell which does not grow with it and must be shed several times a year. During the molt, the lobster is at its most vulnerable and will hide among the rocks and seaweed until its new shell has hardened. They will eat much of their old shell for its lime content, which will be added to the new armor.

At other times the lobster is capable of putting up a good defense. When approached, the lobster immediately begins to wave its antennae to determine the nature and intent of the challenging presence. This rapid movement of the antennae also is a signal to others that danger is near. If an enemy continues to approach, the lobster raises its closed claws, ready to defend itself if necessary. This threat display may discourage small predators, depending on the size of the lobster. If it does not, the lobster will attack by opening its raised claws.

If the stranger is identified as a female lobster, the male will lower and close his claws, allowing her to pass. The courtship display in the breeding season includes a mating dance performed by the male, who moves on the tips of his legs.

A predator itself, the lobster will eat anything it can catch with its claws. As it nears prey it also begins to move its small articulated sliding plates, sometimes called "buccal parts," which serve as its mouth.

On the Pacific Coast, the spiny lobster or saltwater crayfish is found. It lacks the large pincers of its East Coast cousin but uses larger, more developed antennae to strike prey or to ward off enemies. It also grows larger, up to thirty pounds (although they are usually taken when no more than six pounds), but otherwise its behavior is quite similar to the Maine lobster.

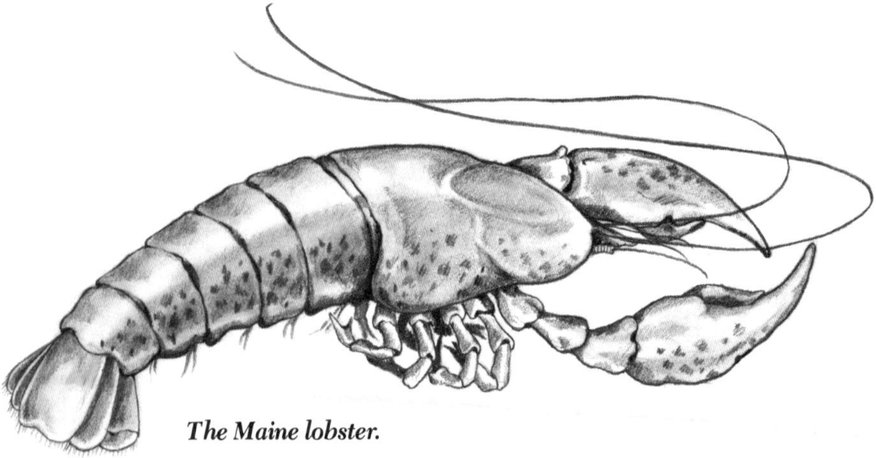

The Maine lobster.

The (Rhesus) Monkey

Because primates so closely resemble humans and because many species exhibit high intelligence and complex communication, we have studied them more closely than other animals. And we have made some remarkable discoveries. In 1969, the year in which the first man set foot on the moon, science experienced a breakthrough which, though much less heralded than the Apollo mission, sent shock waves through the scientific community which are still being felt today. Inspired by Jane Goodall's observations of wild chimpanzees, two American behaviorists, the Gardners, taught a

1. Menace.

3. Intensified fear and menace; cry becomes a loud "rrah."

2. Fear and menace mix; monkey gives a "roo" call.

4. Expression indicates a desire to flee.

5. Unable to flee, monkey makes a loud "eech" cry.

6. Submissive expression as another monkey slowly approaches.

7. Beginnings of sexual arousal.

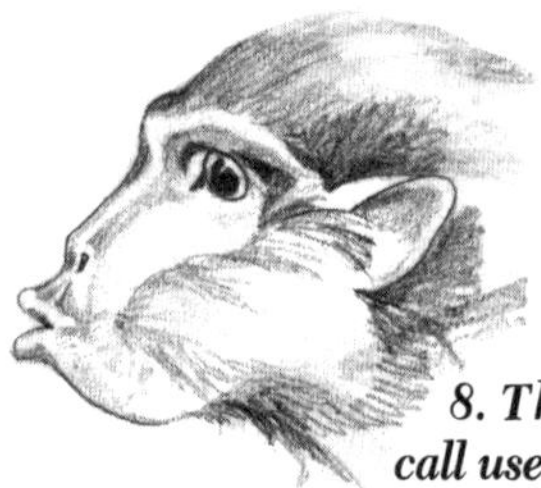

8. The "ou-ou-ou" call used by the young before approaching their mothers.

9. The "ahah" cry before play.

wild-born young female chimpanzee named Washoe to use some 130 recognized signs from the American Sign Language for the Deaf. Since that time, other experimenters have duplicated the Gardners' research and have extended it to include gorillas, which are proving even more intelligent. Though still controversial, the validity of their findings is now widely accepted. Their fascinating work has been documented in films, books and magazine articles.

The rhesus monkey, *Macaca mulatta,* is a native of India and a member of the macaque family. It is widely used in research and experimentation. Rhesus monkeys are somewhat less intelligent than the higher primates, but they too have complex systems of vocal and gestural communication. Here, we will merely hint at their complexity by illustrating several of the more frequent facial expressions.

The Pig

The pig, one of the more intelligent farm animals, is capable of regulating its own diet, keeping itself clean and remaining free of external parasites if raised under the proper conditions. Ironically, its bathing behavior is responsible for its reputation as a messy animal. Like its close cousin and ancestor, the wild boar, the free-roaming domestic pig rolls in mud which, when it dries, is easily brushed off, carrying with it dirt and parasites. Most modern pig farmers do not permit the animals to roam, nor do they provide access to a pond of muddy water similar to the wallowing places of boars, but the pigs will still follow their instincts and try to roll in whatever is available in a valiant, if unsuccessful, attempt to keep clean.

Hierarchy and Aggression

The young piglet's first contact with its peers is in competition for food. Born with sharp little canine teeth, which they will eventually lose, piglets nip each other and make short, extremely shrill cries in their efforts to reach the best nursing site. In this fight for mother's milk is the beginning of social hierarchy. The piglet's rank is established according to its weight and size, with higher ranked piglets accorded exclusive access to the teats which give the most milk. Once the order of feeding has been established, the level of aggression drops considerably.

As the piglets mature, the sharp short milk canines are replaced by permanent teeth, which in the male constitute long and formidable weapons to be used frequently in establishing and maintaining position in the hierarchy. Meeting for the first time, two hogs will sniff and circle, each evaluating the other. The preliminary contact is made by shoving each other, side by side, hair raised along the spine, ears erect, eyes glaring and head raised in an attitude of defiance. Some hogs scratch the ground nervously scattering earth around. The two then face each other shoulder to shoulder, circling continuously, searching for an advantage and trading grunts of aggression—muffled sounds coming from deep in the throat. They also grind their teeth and slap their jaws throughout the fight. As the fight progresses, each hog charges the flank of its opponent. Its head is stretched forward, teeth bared ready to bite the feet, ears and neck—sometimes inflicting serious wounds. If one of the hogs gains the advantage he will bite his adversary's shoulder, but experienced hogs maneuver to avoid such wounds. The fight is over when the weaker hog turns and runs, although the victor will often give chase and attempt a few parting bites.

Fighting between males goes on constantly until the hierarchy has been established, and every newcomer to the group must fight to establish his rank. Even a familiar pig which has been separated for a day or two must fight just to reestablish itself in the hierarchy, and dominant males must, to maintain their hard-won position, continuously reaffirm their power by threats and duels as frequently as once a week. An aggressive glance or a brief grunt are often sufficient to make the subordinate challenger retreat. Some pigs are naturally more aggressive than others, but all pigs, when transported together to market, will exhibit heightened aggressive behavior toward a newcomer and toward each other. To prevent serious harm to either, breeders sometimes tranquilize the animals before shipping them.

Figs. 1–6: Typical fight sequence between two pigs in a struggle for rank.

Sows and Their Young

Sows too establish and maintain a precise social hierarchy, but their fights are not usually as serious as those among males. Rank and authority tend to be determined by the weight and age of the individual sows. Young females sometimes bite—quickly and unpredictably—but older sows are usually too preoccupied with the care of their litters to indulge in combat.

As described above, piglets utter sharp hunger cries to demand food. As they mature the cries grow softer, but the sow will still respond with her usual summoning grunt to reassemble the litter. While the piglets nurse, she grunts softly until she determines they have suckled long enough.

Before she rises, she grunts loudly and the piglets move away from her. If a stranger approaches the litter, she also grunts loudly and at the same time inhales sharply with a sound analogous to a human gasp to warn away the intruder.

Pigs are sociable contact-animals. They have a tendency to form tight groups maintaining close contact—physical, visual and vocal. As a young piglet takes the first steps away from its mother to explore the world beyond, it keeps in touch with her by uttering soft, continuous gasps. When grown animals disperse to feed, they still give a localization cry to keep track of each other. When disturbed suddenly, pigs make a sharp, screeching cry of alarm. In response, other pigs and piglets run off and seek cover.

Courtship

During the breeding season, the male follows the female persistently, stopping often to sniff her. He grunts frequently, grinds his teeth, smacks his jaws together and foams at the mouth. Trailing her by about a yard, the male utters the "courtship song," a series of deep, regular guttural grunts, six to eight per second. After fifteen or twenty grunts he pauses to catch his breath. If the female too has stopped, the courtship song ends.

The ritual of courtship varies from pig to pig and from season to season. Some or all of the above elements may be used in the chase and other behavioral patterns, but one thing is constant. In the pigpen, it is the exclusive prerogative of the dominant male to initiate courtship. He is given first access to the females and, as with other animals, can be most selective in his choice of a mate.

The Pigeon

The wild rock dove, *Columba livia,* is one of the most maligned members of the bird kingdom. It is known to most people by its common name—the pigeon—and is considered an urban pest by many, yet its long history of association with man includes instances of actual heroism. Through the Korean War homing pigeons were used as messengers of last resort by front-line troops, and several of the birds, though badly wounded, got their messages through and saved their human compatriots. Several were decorated for their service, and when the military lofts in Fort Monmouth, New Jersey, were finally closed in the late 1960s, the remaining veteran birds—one aged twenty-one years—were donated to zoos which welcomed these war heroes and cared for them through their final years.

During the Middle Ages the nobility kept enormous pigeon lofts or dovecotes where they raised thousands of birds as an essential source of protein during the long winter months. The privilege of raising and hunting pigeons was reserved for the nobility and was regulated by strict laws. Peasants caught poaching pigeons were severely punished.

Today, especially in Belgium where selective breeding has produced the homing pigeon, they serve as a kind of poor man's race horse, although the owners of champion racing homers are anything but poor. Individual breeding birds are sold at auction for as much as $75,000, and they earn their owners hefty purses in annual long-distance races —the Barcelona-to-Brussels, for example. There is often off-track betting on these races.

Because of the ease of observation, city pigeons make ideal subjects for city-bound naturalists and the rituals and habits of these birds are well worth watching.

Courtship and Territory

Pigeons live in large flocks but even so, they do claim some territory for their own and will make a threat display to discourage intruders. The bird lowers its tail and ruffles the feathers of its neck and body (Figure 1). If the intruder persists, the pigeon crouches slightly, raising a wing on the side opposite the intruder and gives an alarm signal resembling a low growl (Figure 2). In the next stage, the bird flaps its wings in the direction of the intruder (Figure 3) and pecks it while making a menacing "waou" cry. The pecking is a threat rather than a serious attempt to inflict harm.

The same kind of behavior is used by the male pigeon in courting his mate. He will raise his head (Figure 4), ruffle his feathers (particularly his breast feathers) to increase his size (Figure 5) and turn in circles, bowing and cooing before the female (Figure 6).

The pigeon makes a soft guttural cooing and the male will usually string five notes together when courting— "grou-grou-grou-grou-grou." The same series is used in defense of territory.

Courtship near the nest is slightly different. The male inclines his head slightly (Figure 7) and makes a three-noted series of coos, then crouches with tail erect and wings quivering (Figure 8) and continues singing, renewing the three-syllable cry more loudly.

In the wild, pigeons avoid mankind and inhabit high cliffs, building their nests wherever they can find a toehold. In cities and suburbs, pigeons feel at home nesting on ledges, windowsills and in such awkward places as air condi-

Figs. 1–8: The courtship display of the male pigeon.

tioners and on the bas relief on some of our older buildings. The male is responsible for building the nest and shares the work of incubating the eggs once they are laid. Each pair will raise no more than two young at a time, even if more eggs are laid. However, they will breed several times a year and are capable of rapid population increase. Even in the face of growing evidence that they carry diseases to which humans are susceptible—resulting in programs aimed at the eradication of these birds from cities—they continue to multiply, thanks in part to the handouts they receive from humans. Pigeons are easy to capture and when they are fed regularly they will come to expect their benefactor at a certain time and place each day. It is not difficult to coax them onto one's wrist or shoulder and feed them by hand.

Rabbits and Hares

Rabbits and hares both belong to the *Lagomorpha* order (*lagomorph* actually means "shaped like a hare"), but although they share certain physical characteristics, the two species are quite distinct. They will not readily interbreed; their nesting habits are entirely different, and in fact if kept together they will probably end up fighting.

The Rabbit

The ancient Romans tried unsuccessfully to raise and keep rabbits. Their stock kept escaping by energetically burrowing out of the hare gardens in which they were housed. But monks of the Middle Ages did succeed in domesticating the species and raising them for food. Directly descended from these wild ancestors, today's domesticated rabbits come in a variety of sizes and colors, some with long hair, some with smooth fur, some with ears longer than others—all the result of selective breeding but all sharing most behavioral traits with the brown, "garden-variety" cottontail rabbit.

Rabbits are perhaps best known for the frequency with which they breed. The rabbit doe is polyestrus—she has multiple periods of heat during the year and the ova she develops are not released until sexual stimulation occurs.

She can thus breed at almost any time and normally will have four or five litters a year. In captivity, doe and buck will mate as soon as they are put in the same hutch together, but in the wild there is more preliminary activity. The buck begins by pursuing the doe, but only for a few yards. The two stop and eat, or pretend to eat, watching each other. The male uses his tail as a signal, lifting it to expose its white underside. Moving away from the doe, he turns so that she can see this tail signal, then repeatedly advances and withdraws, perhaps three or four times. He may circle her at a trot—sometimes faster—keeping his hindquarters toward her so that she can see the white part of his tail.

A male rabbit raising his tail to show the white underside.

Scent plays a part in rabbit communication. The buck releases an odorous secretion from his anal glands which can be "read" by other rabbits, particularly does. As part of the courtship ritual the buck may spray the doe with a carefully aimed jet of urine. The reaction of the doe to this behavior varies from rabbit to rabbit. Some are indifferent, while others behave as if stimulated by it. The female is capable of the same behavior, shown as an aggressive signal, to dissuade an unwanted buck. The scent is used as a marking device, allowing one rabbit to identify and track another by the scent signature of its urine or feces.

The rituals of courtship—the tail and scent signals—are also part of the rabbit's threat display. Although rabbits are mute, they do use sound when giving an alarm signal. They will thump the ground with their rear paws, usually in the course of running away from danger.

The Hare

One of the major differences between rabbits and hares is the nesting habit of each. Rabbits burrow underground while hares are surface-dwellers, building their nests with leaves. The female hare slaps her large ears together to summon her litter when they have wandered.

The hare has a high-pitched distress signal which it makes when threatened or wounded—a shrill scream that many hunters recognize by its disconcerting resemblance to the cry of a human infant. It warns other hares of danger, but also attracts predators, and hunters have learned to imitate this sound, vocally and with calls, to attract carnivorous game.

The Reindeer

In prehistoric times, the reindeer roamed the immense forests of Europe, furnishing early man with food, clothing, shelter and tools made from its antlers and bones. It was revered and immortalized in cave paintings. Today its range is confined to the Nordic countries—Lapland, Sweden, Denmark and Finland—and the Arctic regions of the Soviet Union. It still provides food and clothing for those who inhabit these cold and forbidding parts of the world and has long been at least partly domesticated, used today to pull sleds over the frozen Arctic.

Conversation

Wild and tamed reindeer have a similar and varied vocal repertory. To call their young, mothers use a summoning cry which consists of a series of very short (less than one-third of a second each) "ha-ha-ha" sounds repeated at intervals of not more than a minute. The response of the young is similar.

During courtship, the male's call is a series of raucous and rapid rattles made by exhaling air. He stretches his head horizontally, his mouth half open. When near a female, he claps his tongue against his palate, making a sound audible only to a female a few yards away.

Occasionally two males fight, usually during the breeding season. Fighting is preceded by a threat display during which both animals paw the ground with their front feet and rub and butt their antlers against trees and tufts of grass. As with other horned or antlered mammals, the reindeer's antlers play an important symbolic role. Often its rank in the herd can be directly correlated to the size of its antlers, and if it loses one or both its place will be considerably modified. Female reindeer are unusual in that they are the only antler-bearing females of the family of *Cervidae,* or deerlike mammals. Both sexes shed their antlers annually, the males dropping theirs earlier than the females. During that time, and only temporarily, the females have the greater authority in the herd.

Reindeer buck pawing the ground.

Reindeer and Men

The breeders in Lapland use a whole series of calls to summon their herds, which are only semidomesticated. Many of these calls resemble those made by the reindeer, but different calls are made in the different regions in which these animals are found. For example, the Swedish herders use short "eu-eu-eu" sounds, while in Jutland, on the mainland of Denmark, reindeer respond to short "kik-kuk-kiu" cries followed by a long "tsy" and a very brief "a." Despite a certain resemblance to reindeer's cries, these calls are arbitrarily human and must be learned by the reindeer.

The Robin

In those regions where the robin breeds and spends the summer months, it is considered a harbinger of spring—one of the first birds to return after the long winter months. Actually many robins stay in the north year round, although during the winter they retreat from their usual haunts into forests and cedar bogs and are rarely seen until spring.

Communication

The robin uses six basic sound signals. Two of these are alarm cries warning others of the presence of danger—a single "tic" sound or a series of repeated "tics" given while the bird is in motion or in flight. Other cries signify distress, aggression and the female's search for a mate. Young robins make a contact cry that sounds very much like "zeep."

More complex vocalizations are difficult to transcribe. The robin's song is melodious and conveys a great deal of information about the bird's identity and territory. It warns away or intimidates intruders and it also invites a mate. The song is punctuated by pauses to allow either intruder or mate to respond.

If an encroaching robin does not leave, the resident robin makes a threat display, ruffling its feathers and singing an altered song. These measures will usually cause the intruder to withdraw without an actual physical confrontation.

Top: Robin feeding its mate as part of the courtship display. Middle: Threat display. Bottom: Position of alert.

Robins and Humans

The robin readily tolerates human presence and is easily tamed. It will return regularly to a given location to be fed by hand. Robins have been observed following a few steps behind a gardener raking leaves, feeding on the worms and bugs uncovered by the rake.

One must be careful not to approximate the cries of distress, alarm or aggression when the object is to befriend a robin, and after much patient and persistent practice it is possible to distinguish and imitate the cry of a female robin. Bird-watchers have actually attracted male robins this way.

Scientists have provoked the threat display described above by gluing a few red feathers to a stick and introducing it into the robin's territory. The robin responded to the decoy—and to its red color—as it would to a rival bird. In another experiment, the boundaries of an individual robin's territory were determined by propping up a large mirror on the ground at varying distances from the bird. Seeing its own image, the robin responded as if another robin were nearby. If the mirror was inside the robin's territory, it would approach and threaten the image. If the mirror was outside the bird's territory the bird would perhaps investigate but would not be bothered by the presence of the other robin. The territorial boundaries were determined by moving the mirror to different locations and observing the bird's reaction each time.

A robin responds to its reflection, revealing the limits of its territory.

The Stickleback

The stickleback is a small pugnacious fish with a distinctive courtship and mating ritual which has long fascinated scientists. Because it breeds readily in the aquarium, it has been an ideal subject of observation.

In the breeding season, usually the spring, the abdomen of the male stickleback turns reddish in color. This attracts females and also enables males to identify other males who intrude in their territory. The stickleback is strongly territorial and the inexperienced male will attack fish of any color until its first confrontations with rival male sticklebacks. After that, it will concentrate its efforts, but only insofar as it will still attack fish or even inanimate objects which are red.

Once the male has established his territory, he constructs a small tunnel-like nest of weeds in the center of it. Then, by swimming on his side, he exposes his red belly and attracts the female. With a distinctive zigzag dance, he leads her to the nest and there he points to the entrance with his nose. Before following, the female arches her head and tail, then enters the nest.

The male nudges her from behind to stimulate her to lay eggs, which she deposits in the nest. The male takes her place, spraying the eggs with a layer of sperm, then drives off the female whose part in the cycle is finished. The eggs are tended and guarded by the male, whose red color

The male stickleback attracts and leads the female to the nest.

quickly fades to make him less conspicuous among the weeds. He tries to keep the young hatchlings near the nest to protect them from predators, but eventually all the young sticklebacks disperse, and only a minority will survive to maturity.

The Stork

In the United States there is only one species of stork, the wood stork, *Mycteria americana,* which is found from South Carolina to Florida and west to Texas. It shares many characteristics, physical and behavioral, with its more famous European cousins, *Ciconia alba*—the white storks of legend. They nest as far north as the Netherlands, where their return each spring is eagerly awaited in towns which consider their presence a symbol of good luck.

Unfortunately, in recent years the birds themselves have needed more than luck to survive the predation of man in their winter feeding grounds in Africa and the loss of their habitat in the north where much of the marshland that provided them with food has been drained. Many birds annually make poor choices of nesting sites and assemble their piles of sticks and debris on top of electric pylons only to fall victim eventually to the high tension wires. Other birds have been killed by ingesting high levels of pesticides such as parathion, used to kill grasshoppers. Some scientists have found evidence that the humidity in Europe is higher than it once was, and it is thought that this too might be affecting the birds. Whatever the reason, as with the whooping crane of the United States, these pressures have pushed the species to the brink of extinction, and in 1977 fewer than fifteen breeding pairs returned to their once-populous habitat in the Strasbourg region of France. In an effort to stop the dwindling of the species a breeding program has been established known as the Center for Reclamation of the Stork. Rather than allowing the young storks to migrate to Africa where they stand a good chance of dying in any number of ways, they are kept and raised to maturity in a nursery at Kintzheim-Hunewihr in France. Whether or not the program will succeed over the long term remains to be seen.

The white stork.

Communication

The long-necked stork lacks muscles in its voice box and is therefore almost completely mute, except for a loud hissing it makes as a warning against intruders. The most dramatic behavior to observe is the courtship display during the breeding season. In the spring when the birds have returned from their wintering-over areas, pairs establish themselves in nests, often built on man-made nesting platforms overlaid by successive generations with sticks and debris. They will roost either in colonies or single pairs. The courtship involves a mutual display of loud bill clattering—similar to the fencing of the gannet—and a ritual in which the two birds perch on the nest, each with its body leaning forward, head back and wings beating in irregular rhythm. The male executes aerial acrobatics around the nest. After copulation occurs, both sexes take turns incubating the eggs which hatch in approximately thirty-three days. When the chicks are born, the parents also take turns searching for food.

A parent returns with food for the hatchlings.

Before alighting on the nest, the returning bird claps its beak several times to announce its arrival. Without this signal, the bird sitting on the nest may attack, using its bill as the formidable weapon it is. Occasionally a mate has received a fatal blow by mistake.

To identify itself to its mate, the returning stork claps its bill loudly before landing.

The white stork is perhaps the most adapted to the presence of man. On the breeding farm, individuals have grown quite tame, allowing themselves to be petted and approaching visitors who picnic in the park. They have even been observed taking food from a visitor's plate.

Afterword

It is hoped that the reader, having been introduced here to the elaborate and manifold channels by which animals communicate, will want to pursue further observations of animal language and body language through which we humans may come to understand that broad range of animal behavior whose messages we are just beginning to decode.